G. Marshall

Head of Mathematics Department
Ellers High School, Doncaster

A World of Mathematics

MATHS ON TARGET 1

Nelson

ISBN 0 17 431237 7

Thomas Nelson and Sons Ltd.
Lincoln Way Windmill Road
Sunbury-on-Thames Middlesex TW16 7HP
P.O. Box 73146 Nairobi Kenya

Thomas Nelson (Australia) Ltd.
19-39 Jeffcott Street West Melbourne Victoria 3003

Thomas Nelson and Sons (Canada) Ltd.
81 Curlew Drive Don Mills Ontario

Thomas Nelson (Nigeria) Ltd.
8 Ilupeju Bypass PMB 1303 Ikeja Lagos

Design by Gerry Watt

Printed in Hong Kong

Acknowledgements

Acknowledgements are due to the following for photographs
reproduced in this book:

British Gas Corporation, 62;

Central Office of Information (Crown Copyright), 22 (left & top), 24, 78,
93 (top), 96 (both);

Patrick Eagar, 18 (top-left & top-middle), 19 (right);

Henry Grant, 3 (both), 4;

Hoover Limited, 34 (top), 35, 38 (foot & top-left), 39 (right);

I.B.M. Ltd., 97, 102;

London Photo Agency, 16 (top), 17, 21 (left);

Abbey National, 53 (right);

The Post Office, 11 (top-left, top-right & foot), 12, 13, 51;

Tube Investments Limited, 65;

Youth Hostels Association, 40 (top-middle, bottom-right & bottom-left),
42 (right), 43 (right), 44;

Barnabys Picture Library, 22 (both), 31;

Trustee Savings Bank, 49;

Cyril Bernard, 67.

Contents

1

Earning a living

What are the important things to consider when you are looking for your first job? Here are a few suggestions.

* Will I like the work?
* Will I be strong enough?
* What are the hours and holidays?
* Will I be allowed time off to study?

Try to think of some other things that are important.

Someone certainly will have mentioned **pay**. You will need to know how your wages are worked out and you will be wise to check them every week. In this chapter we are going to look at some of the more common ways of working out wages.

Assignments

Payment by the hour

Some workers, such as lorry drivers and men in building trades, are paid by the hour. Their wages depend on:

* the number of **hours** they work in a week.
* their **rate of pay** for each hour.

1

A Copy this table carefully and find the total number of hours worked by each man in the week.

	Slater	Carpenter	Driver	Baker	Woodman	Smith	Webster
Monday	8	6	7	8	5	6	8
Tuesday	8	6	8	8	6	7	8
Wednesday	8	7	7	6	7	7	7
Thursday	8	8	5	4	4	8	6
Friday	8	5	5	5	6	7	8
Total							

B Each trade has its own agreed rate of pay for an hour's work.

Here is an example.

Mr. Coleman works a total of 40 hours in a week and his rate of pay is 80p an hour. His wage for the week is:

$$80p \times 40 = 3200p$$
$$= £32.00$$

Use the ready reckoner on page 3 to work out the wages for each man in the list below.

	Name	Hours	Rate
1	A. Taylor	40	70p
2	I. Tanner	40	55p
3	E. Glazier	37	80p
4	R. Cooper	39	75p
5	Y. Chapman	40	100p
6	A. Glover	34	65p
7	C. Fisher	38	95p
8	G. G. Farrier	33	55p
9	T. Farmer	32	75p
10	C. Docker	35	90p
11	T. Brewer	34	55p
12	R. Teacher	40	90p
13	A. Sherrif	31	50p
14	B. Constable	33	85p
15	L. Sargeant	36	65p
16	I. Page	39	50p
17	A. Cook	40	70p
18	R. Butcher	38	95p
19	L. Draper	40	60p
20	B. Chandler	34	75p
21	P. Turner	40	85p
22	N. Shooter	31	70p
23	T. Fryer	35	85p
24	D. Clark	32	95p

Hours	Rate p										
	50	55	60	65	70	75	80	85	90	95	100
31	1550	1705	1860	2015	2170	2325	2480	2635	2790	2945	3100
32	1600	1760	1920	2080	2240	2400	2560	2720	2880	3040	3200
33	1650	1815	1980	2145	2310	2475	2640	2805	2970	3135	3300
34	1700	1870	2040	2210	2380	2550	2720	2890	3060	3230	3400
35	1750	1925	2100	2275	2450	2625	2800	2975	3150	3325	3500
36	1800	1980	2160	2340	2520	2700	2880	3060	3240	3420	3600
37	1850	2035	2220	2405	2590	2775	2960	3145	3330	3515	3700
38	1900	2090	2280	2470	2660	2850	3040	3230	3420	3610	3800
39	1950	2145	2340	2535	2730	2925	3120	3315	3510	3705	3900
40	2000	2200	2400	2600	2800	3000	3200	3400	3600	3800	4000

Piece work: payment for work done

In many factories workers are paid according to the amount of work they do; for example the number of parts of a machine they make or assemble.

Their wages depend on:

* the number of parts (pieces) they make or assemble in a week,
* the rate of pay for each part.

The piecework rates are usually agreed between the firm and the trades union representing the workers.

Here is an example.

Piecework rate: £0.25 per 1000 pieces
Number of pieces (parts) completed in the week: 100 000

$$\text{Wages: } \frac{100\,000}{1000} \times £0.25 = 100 \times £0.25$$

$$= £25$$

Find the week's wages for each of these men if the guaranteed minimum wage is £25.

Week ending 25th October

	Name	Trade	Piecework rate	Work completed (pieces)
1	L. Phillips	buffer	£0.25 per 100	12 000
2	M. Oborne	sprayer	£0.30 per 100	11 000
3	L. Mason	packer	£0.24 per 100	8 000
4	C. Potts	driller	£0.28 per 100	10 000
5	P. Hughes	grinder	£0.32 per 100	7 000
6	S. Brooks	miller	£0.55 per 100	6 000
7	I. Jones	fitter	£0.62 per 100	4 000
8	S. Pearson	turner	£0.44 per 100	6 500

C Work out the weekly wages for each of these examples.

	Name	Piecework rate	Work completed (pieces)
1	W. Bell	£0.30 per 100	8 000
2	J. Mills	£0.30 per 100	9 000
3	R. Delaney	£0.30 per 100	10 000
4	D. Coates	£0.35 per 100	6 000
5	D. Cook	£0.35 per 100	7 000
6	D. Minogue	£0.35 per 100	8 000
7	V. Holmes	£0.35 per 1000	80 000
8	K. Adams	£0.40 per 1000	80 000
9	N. Smith	£0.45 per 1000	60 000
10	L. Peters	£0.15 per 10	1 200

D In some factories the workers are guaranteed a **minimum wage**. For example, if the minimum wage is £20 and a man earns only £19 he is paid £20 in wages.

Basic wage plus commission

Many shop assistants are paid a basic weekly wage plus an extra payment called **commission**. Commission is usually worked out as a percentage of the value of the goods that the assistant sells in a week.

Here is an example:

'Angela Taylor receives a basic wage of £8 plus 5% commission on the goods she sells. If her sales amount to £60 in one week what will her total wages be?'

Commission 5% of £60	= £3.00
Basic wage	= £8.00
Total wages	= £11.00

Notice that **5% commission** is the same as **5p in the £1 commission.**

E 1 Copy and complete this table to find the total wage for each of the ten girls.

	Total sales	Rate of commission	Commission earned	Basic wage	Total wages
I. Sellers	£70	4%	£2.80	£9.00	£11.80
A. Packer	£85	2%	£1.70	£10.00	
J. Smith	£100	3%		£8.00	
M. Weaver	£90	4p in the £	£3.60	£7.00	
R. Davies	£120	3p in the £		£11.00	
L. Jones	£85	5%		£7.00	
C. Brown	£200	2p in the £		£8.50	
M. Summers	£120	3%		£9.50	
O. Barker	£135	4%		£6.25	
E. Firth	£300	1½p in the £		£6.50	

2 Christine Rennison works in a Record Centre. She receives a basic wage of £5, 5% commission on weekly sales up to £50 and 10% commission on sales over £50. How much did she earn in a week when her sales totalled £80?

3 Mr. Sewell, a commercial traveller, earns a basic wage of £20 plus commission of 3p in the £ on all weekly sales over £100. How much does he earn in the following weeks?

Week ending	Dec. 5th	Dec. 12th	Dec. 19th	Dec. 26th	Jan. 2nd
Sales	£1000	£1200	£800	£400	£350

4 Caroline Brown sells cosmetics. She is paid a basic wage of £7.50 plus 3p in the £ on weekly sales up to £100 and 5p in the £ on sales over £100. How much does she earn in a week when her sales amount to £120?

2

Maps and plans

Here is a plan of a desk top.

Here is a map of England, Scotland and Wales.

The scale of a map is sometimes shown by a line.

In this example one centimetre on the map represents a distance of one kilometre.

> 1 cm represents 1 km
>
> 1 cm represents 1000 m
>
> 1 cm represents 100 000 cm

From the line above we can see that an actual distance is **one hundred thousand times** longer than the corresponding distance on the map. We can therefore write the scale in the form:

> 1/100 000   **or**   1:100 000

This is known as the **representative fraction (R.F.)** of the map.

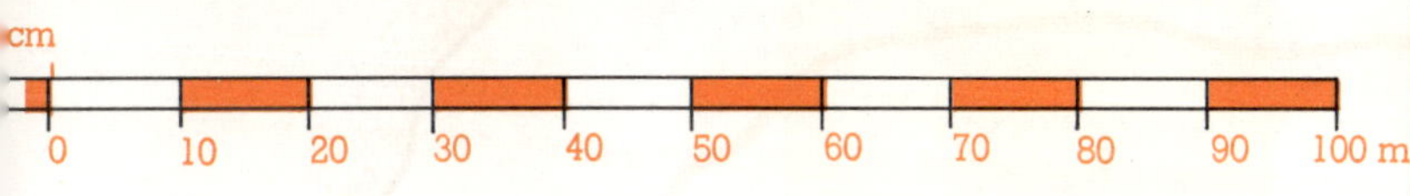

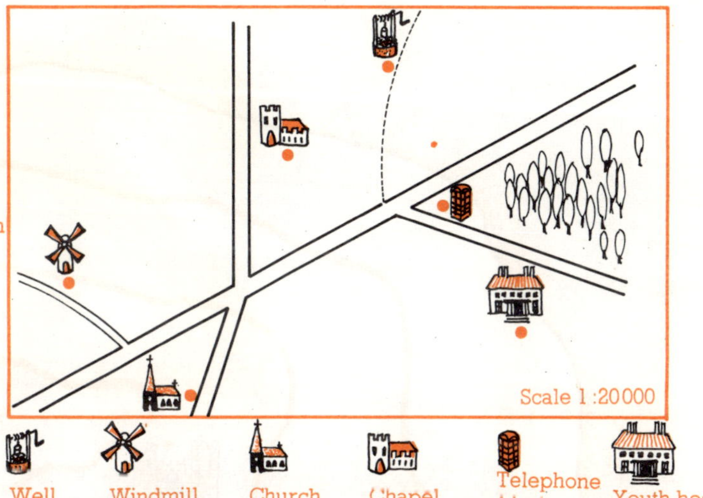

A

1 The above line gives the scale of the plan of a building site.

Copy and complete the following statements and find the **R.F.** (representative fraction) for the plan.

1 cm represents 10 m

1 cm represents ▨▨▨▨▨ cm

The **R.F.** of the plan is 1: ▨▨▨▨▨

2 Find the distances represented by the following measurements on the scale at the top of the page.

a	1 cm	f	0.2 cm (2 mm)	k	3.8 cm
b	2 cm	g	0.6 cm	l	15 cm
c	4 cm	h	1.4 cm	m	20.5 cm
d	5 cm	i	0.5 cm	n	25.1 cm
e	10 cm	j	2.5 cm	o	32.3 cm

3 If the **R.F.** of a plan is 1:100 what actual distances are represented by:

a	1 cm	f	15 cm	k	1.5 cm
b	3 cm	g	18 cm	l	4.8 cm
c	5 cm	h	23 cm	m	14.8 cm
d	7 cm	i	0.1 cm	n	22.2 cm
e	10 cm	j	0.3 cm	o	27.9 cm

B

By measuring and using the **R.F.** find the direct distances between:

1 the well and the church

2 the church and the chapel

3 the youth hostel and the telephone

4 the windmill and the church

5 the chapel and the well

6 the telephone and the church

7 the youth hostel and the chapel

8 the well and the windmill

9 the church and the youth hostel

10 the windmill and the chapel

C

We can find the distance between two places on a map by referring to the scale. Suppose we want to find the height of a particular place? If we were walking or cycling in the country we should certainly be interested in whether a place was at a height of 700 m or at sea level.

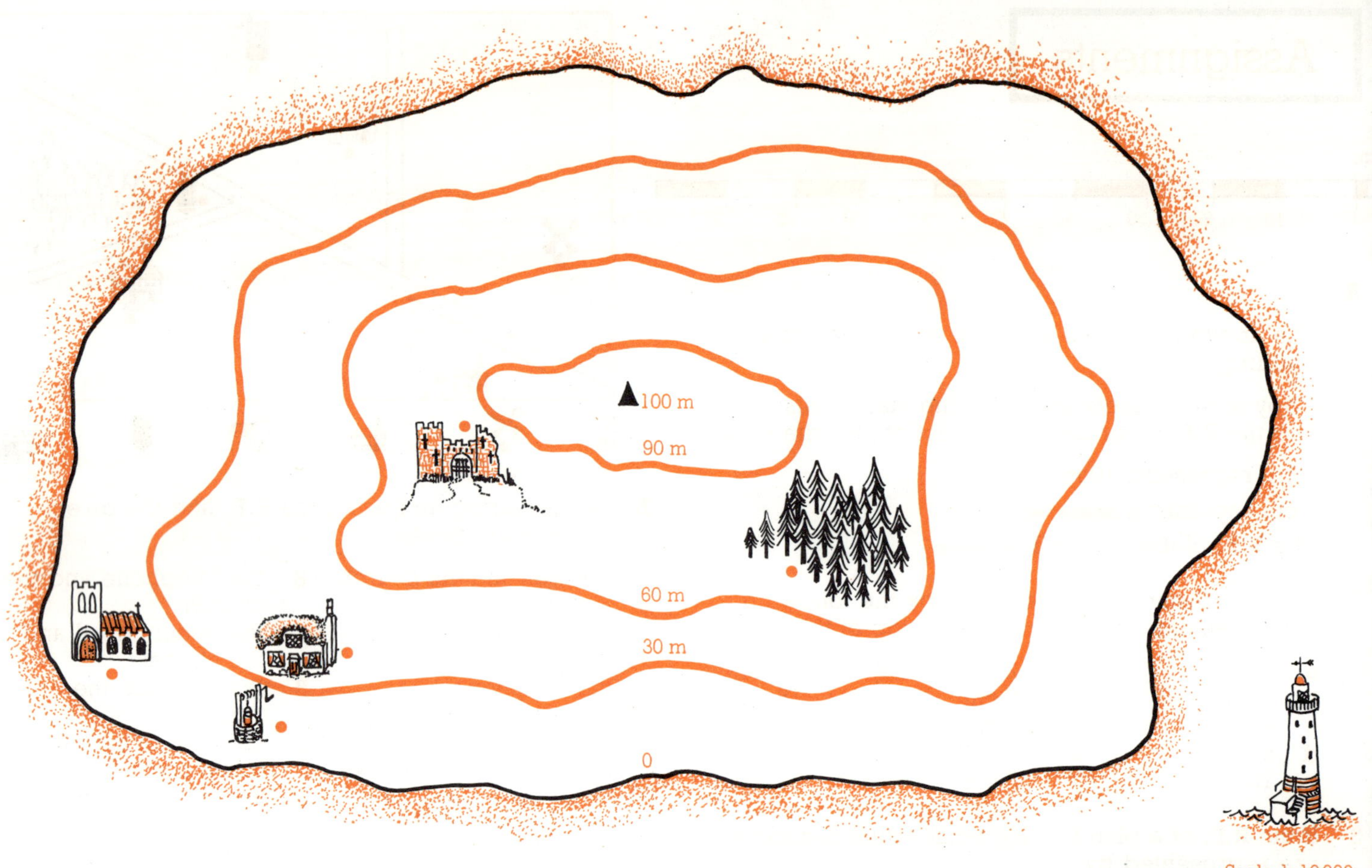

We can, in fact, find the approximate height of a place by looking at the **contour lines** on a map.

Above is the map of an island. Along the line marked 30 m each point is 30 metres above sea level, along the line marked 60 m each point is 60 metres above sea level and so on. The highest point, marked ▲ 100 m, is called a **spot height**.

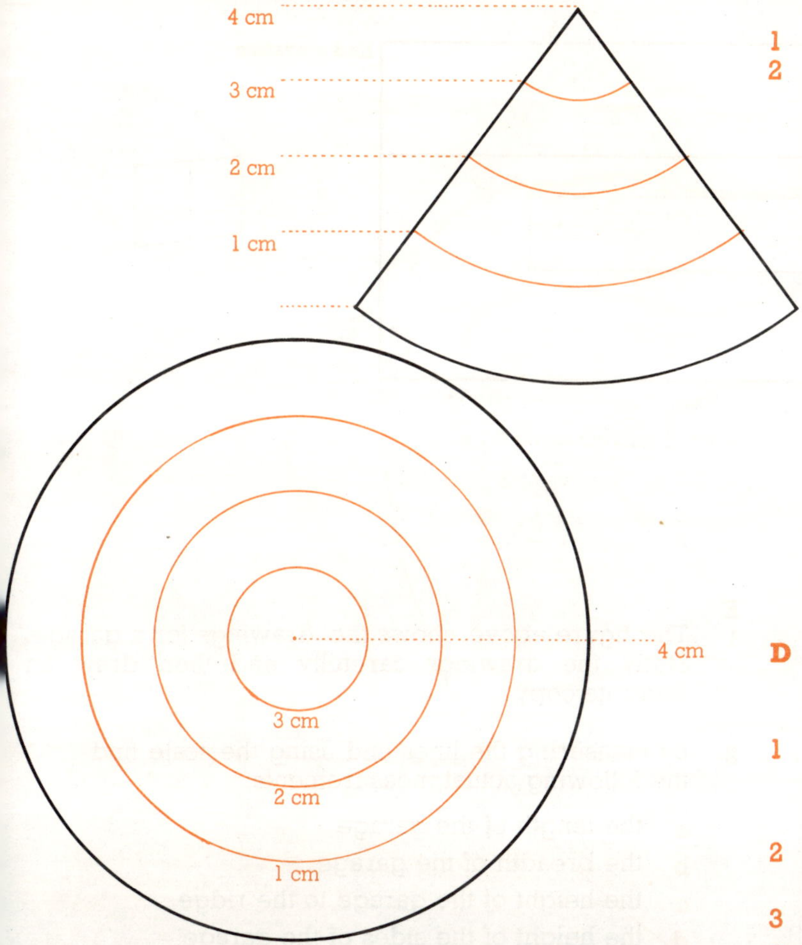

The figure above shows two views of the same cone, first from the side and then from directly above.

The coloured circles drawn round the cone are at vertical heights of 1 cm, 2 cm and 3 cm, and you will see on the plan view that they appear as contour lines.

1 Copy the two views of the cone shown opposite.
2 Copy the plan of the geometric solid shown below, say what the solid is, and then sketch two different side views of it.

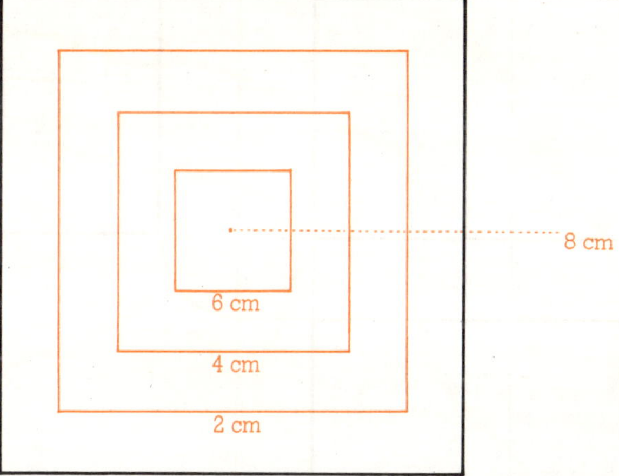

D Look back at the map of the island on page 8 and then answer the following questions. Your answers will only be approximate.

1 Find the height above sea level of:
   a the church   **b** the well   **c** the wood
   d the ancient fort   **e** the cottage.

2 How many metres is the ancient fort below the top of the hill?

3 How much higher up is the cottage than the church?

4 How many metres would you have to climb from the cottage to the top of the hill?

5 How far is the wood from the top of the hill?

6 How far is the lighthouse from the shore?

7 How far is the cottage from the well?

9

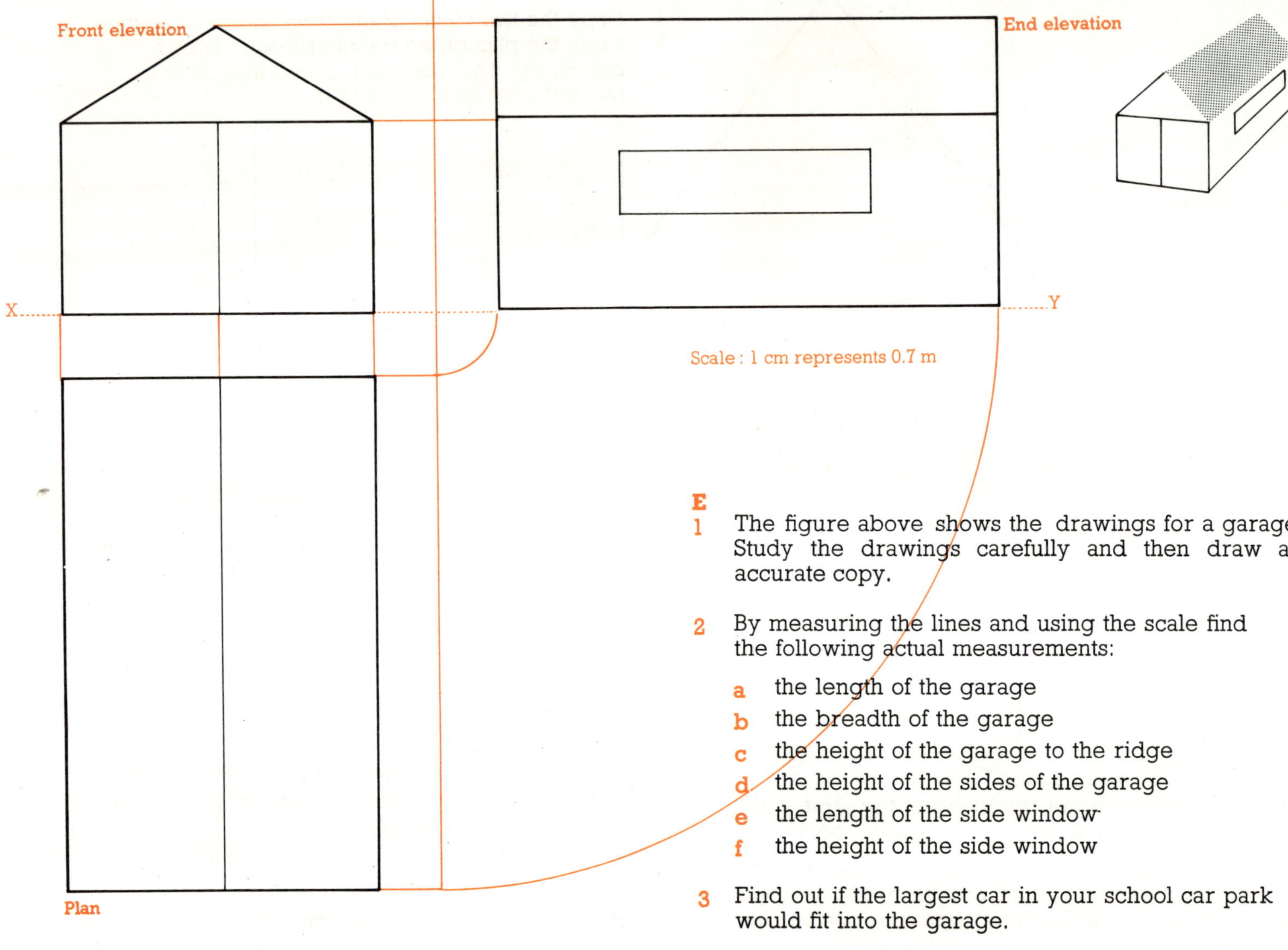

E

1   The figure above shows the drawings for a garage. Study the drawings carefully and then draw an accurate copy.

2   By measuring the lines and using the scale find the following actual measurements:

    a   the length of the garage

    b   the breadth of the garage

    c   the height of the garage to the ridge

    d   the height of the sides of the garage

    e   the length of the side window

    f   the height of the side window

3   Find out if the largest car in your school car park would fit into the garage.

3

The post office

When we hear mention of the Post Office we probably think about buying stamps and posting letters. In fact the Post Office provides a wide variety of services which are essential to the many shops, offices, and factories throughout the country.

Since the Post Office rates change from time to time, you will get the most benefit from the assignments if you first make a collection of up-to-date leaflets.

Weight not over	First Class	Second Class
360g	14p	10p
420g	16p	11½p
480g	18p	13p
720g	27p	18½p (maximum)
960g	36p	—

Assignments

A Post–Office services

Copy and complete these sentences which give a list of the main services provided by the Post Office.

1 The Post Office provides a service for the delivery of ▭▭▭▭ and ▭▭▭▭

2 Urgent messages can be sent either by using the ▭▭▭▭ or sending a ▭▭▭▭

3 ▭▭▭▭ Orders and ▭▭▭▭ Orders provide safe ways of sending money through the post.

4 The Post Office runs the ▭▭▭▭ Bank for the benefit of ordinary people.

5 Post Offices sell licences for ▭▭▭▭ and ▭▭▭▭

6 National G ▭▭▭▭ is a simple and cheap banking service run by the Post Office.

B Inland Postage Rates

The following table is similar to ones prepared by the Post Office. You can use the table for the questions.

Rates of postage for Inland Letters

Weight not over	First Class	Second Class
60g	3½p	3p
120g	5p	4p
180g	8p	5½p
240g	10p	7p
300g	12p	8½p

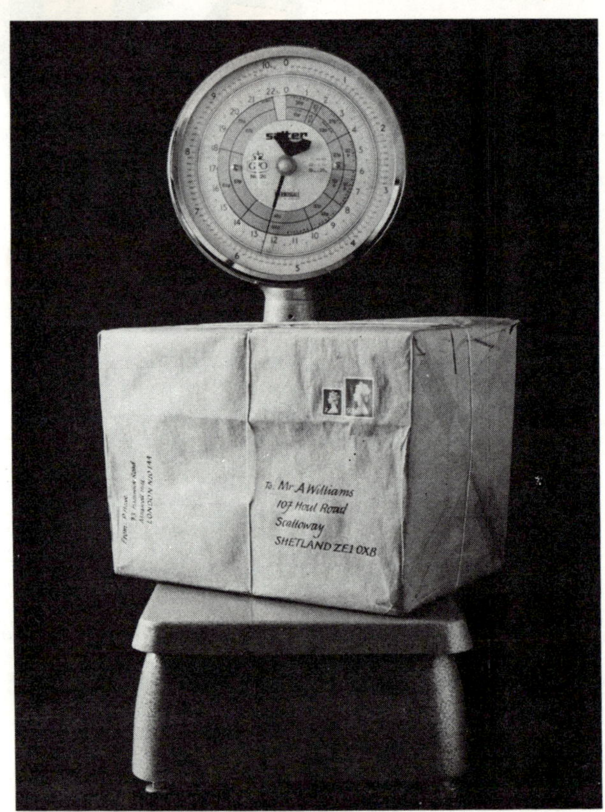

Find the cost of sending the following items by letter post:

1 10 Christmas Cards, each under 60g, by second-class mail.

2 5 letters, each under 60g, by first-class mail.

3 2 packages, each weighing 200g, by first-class mail.

4 A dozen packages, each weighing 500g, by second-class mail.

5 6 calendars, each weighing 250g, by first-class mail.

6 3 packages, each weighing 640g, by first-class mail.

7 100 letters, each weighing less than 60g, by second-class mail.

8 4 packages, each weighing 350g, by second-class mail.

9 1000 circulars, each less than 60g, by second-class mail.

10 Half a dozen birthday cards, each under 60g, by first-class mail.

C Make a model of the largest package which may be sent by letter post.

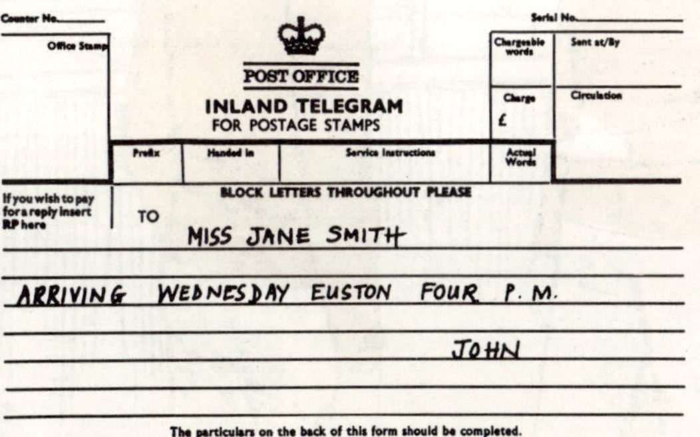

Urgent or special messages can be sent by telegram. Here are some of the important things to know about telegrams.

* They can be handed in at a post office and they should be written in **block** capitals.

* They can be dictated over the telephone.

* The charge depends on the number of words, including the address.

* The words in the address after the name of the road or street are counted as one word.

* A number (for example, 214) is counted as one word.

* There is an extra charge for greetings telegrams.

D Find out the present-day rates for telegrams, either by telephone or by calling at a post office, and then work out the costs of the following telegrams.

kind of telegram	number of words
1 ordinary	12
2 greetings standard	12
3 greetings de luxe	12
4 ordinary	15
5 ordinary	22
6 ordinary	17
7 greetings standard	14
8 greetings de luxe	16
9 ordinary	23
10 ordinary	30
11 greetings de luxe	21
12 greetings standard	19

E Work out the charge for **each** of the following telegrams.

1 JOHNSON 40 GREEN LANE LIVERPOOL GOODS DESPATCHED TODAY PRICE AS AGREED PARKINSONS MANCHESTER

2 CLARK 34 SURREY ROAD CAVERSHAM READING PITCH WATERLOGGED MATCH CANCELLED WRITING TODAY FOR NEW FIXTURE FLETCHER WATFORD

3 KEITH HUNT 99 GRENVILLE ROAD WARMSWORTH DONCASTER MANY CONGRATULATIONS ON YOUR EXAMINATION SUCCESS AUNT PAULINE AND UNCLE GEORGE

4 MR THORNYCROFT 29 QUEENS DRIVE PETERBOROUGH ACCOMMODATION BOOKED AS REQUESTED EXPECT YOU SATURDAY MRS JONES NEWQUAY

5 MR AND MRS LOVEJOY REGENT HOTEL LANDSDOWNE ROAD YORK CONGRATULATIONS ON A FIRST CLASS MATCH AND BEST WISHES FOR THE FUTURE FROM ALL YOUR TEAM MATES AND FRIENDS AT THE ROVERS

6 CHAPMAN 9 EVERINGHAM STREET SHEFFIELD MONDAY LAST DAY FOR TOURNAMENT ENTRIES FULL NAME AND AGE REQUIRED FOR ALL COMPETITORS IMMEDIATE REPLY ESSENTIAL SUMMERS

F 1 Write out two telegrams, one an ordinary telegram and the other a greetings telegram.

2 Work out the costs for the two telegrams you have written.

Sending money by post

Try to answer these questions before you start the next assignment.

a Why is it unwise to send pound notes in a letter through the post?

b What are the advantages of sending money by postal order and by money order.

c What is meant by 'poundage'?

d What are the present charges on postal orders and money orders?

e What is the reason for having both postal orders and money orders?

G Here are the values of postal orders which can be bought. Copy the table and fill in the poundage when you have found the information from a post office.

Value	5p	7½p	10p	12½p	15p	17½p	20p	22½p	25p
Poundage									

Value	30p	35p	40p	45p	50p	55p	60p	65p	70p
Poundage									

Value	75p	80p	95p	£1	£3	£4	£5		
Poundage									

Find the total cost, including poundage, of sending the following:

1 Two postal orders costing 5p

2 A postal order costing 15p

3 Four postal orders costing £1

4 Ten postal orders for 22½p?

5 Postal orders for £1.50

6 Postal orders for £2.25

7 Postal orders for £3.05

8 Three postal orders for 15p

9 Postal orders for £3.65

10 Postal orders costing £4.55

11 Ten postal orders costing £3

12 Two postal orders costing 17½p

13 Postal orders costing 58p

14 Postal orders costing £1.36

15 Postal orders costing £5.57½

16 Five postal orders costing 80p

4

Sports and pastimes

DIVISION 1

	HOME				Goals		AWAY			Goals		
	P	W	D	L	F	A	W	D	L	F	A	PTS
LEEDS	41	12	8	1	38	18	11	6	3	27	13	60
LIVERPOOL	39	18	2	0	34	10	4	9	6	15	17	55
IPSWICH	41	10	7	3	38	20	8	4	9	29	37	47
DERBY	41	12	7	1	38	16	4	7	10	12	26	46
BURNLEY	40	9	8	2	25	15	6	5	10	27	37	43
STOKE	40	12	6	2	38	15	1	10	9	14	27	42
Q.P.R.	39	8	10	2	30	16	5	6	8	25	30	42
EVERTON	40	11	7	1	28	11	4	5	12	21	34	42
LEICESTER	39	9	7	4	32	17	3	9	7	16	22	40
WOLVES	40	10	6	4	29	18	2	9	9	19	29	39
ARSENAL	39	9	6	5	22	15	4	6	9	22	32	38
SHEFF. UTD	41	7	7	7	25	22	6	5	9	18	27	38

Few people regard mathematics as a 'game' to be played in their leisure time but it is surprising how often mathematics plays a part in sports and pastimes.

Here are some examples.

Assignments

A Football

1 If eight teams are to play each other **once** in a **league** competition how many games will be played altogether?
Study the following table, copy the part that is given, and then continue the table for six, seven and eight teams.

number of teams	teams	matches	calculation
3	A	AB AC 2	
	B	BC 1	$\dfrac{3 \times 2}{2} = \dfrac{6}{2} = 3$
	C	$\overline{3}$	
4	A	AB AC AD 3	
	B	BC BD 2	$\dfrac{4 \times 3}{2} = \dfrac{12}{2} = 6$
	C	CD 1	
	D	$\overline{6}$	
5	A	AB AC AD AE 4	
	B	BC BD BE 3	$\dfrac{5 \times 4}{2} = \dfrac{20}{2} = 10$
	C	CD CE 2	
	D	DE 1	
	E	$\overline{10}$	

2 Find, by calculation, the total number of matches played in leagues containing **a** 12 teams **b** 15 teams **c** 20 teams and **d** 22 teams, if each team plays all the other teams **twice**.

3 If there are eight teams in a **knockout** competition how many matches will be played altogether?

Once again study and copy the table given and then continue for six, seven and eight teams. Each coloured line denotes one match, and we assume that the top team of the pair wins the match.

4 Give a rule to find the total number of matches played in a knockout competition. Can you find a simple reason why this rule works?

number of teams	teams	matches	total number of matches
3	A B C	A B → A C → A → A	2
4	A B C D	A B → A C → C → A D	3
5	A B C D E	A B → A C → C → A D → E → A E	4

B Cricket

A cricketer is often judged by his average;
his average runs per innings if he is a batsman,
his average number of runs per wicket if he is
a bowler.

| A batting average = | total number of runs scored / total number of times out |

Example

A batsman scored: 12, 16, 24 not out, 18, 26, and 14.

His batting average is:

$$\frac{12 + 16 + 24 + 18 + 26 + 14}{5} = \frac{110}{5} = 22$$

| A bowling average = | total number of runs scored off the bowling / total number of wickets taken |

Example

A bowler took: 5 wickets for 29 runs, 3 wickets for
25 runs, and 2 wickets for 31 runs.

His bowling average is: $\dfrac{29 + 25 + 31}{5 + 3 + 2} = \dfrac{85}{10} = 8.5$

Assignments **1** and **2** give practice in the operations
needed to find an average.

1 The number line and the examples show how to
correct a number to one decimal place,
that is to the nearest tenth.

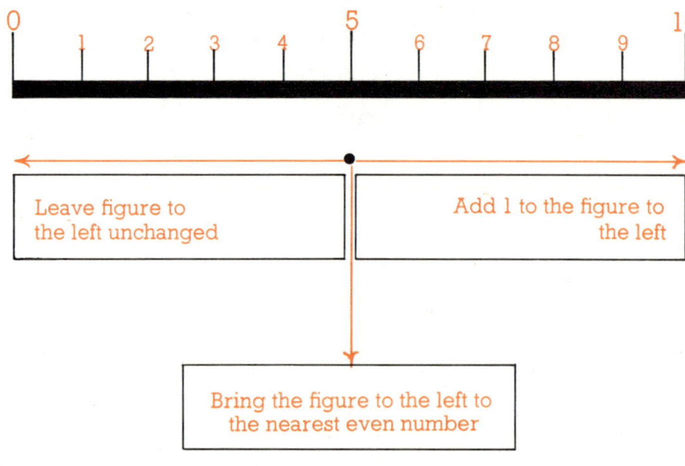

Examples

1.53 → 1.5	1.57 → 1.6	1.55 → 1.6
24.22 → 24.2	24.29 → 24.3	24.99 → 25.0

Correct the following numbers to the nearest tenth.

a	0.11	f	2.03	k	10.78	p	20.98
b	0.14	g	2.09	l	10.87	q	32.55
c	0.16	h	2.04	m	12.39	r	49.66
d	0.18	i	2.08	n	12.93	s	31.09
e	0.15	j	2.05	o	15.56	t	99.99

2 The operations: **addition, subtraction**, and **multiplication** always have an exact answer.

Division is the odd man out, most examples do not have an exact answer.

The following examples are worked out to the second decimal place (hundredths) and are then corrected to one decimal place (the nearest tenth).

```
    5.83 → 5.8        5.66 → 5.7
  6)35.00           9)51.00
    30                45
     5 0               6 0
     4 8               5 4
       20                60
       18                54
```

Work out the following examples correct to one decimal place.

a	15 ÷ 6	f	42 ÷ 9	k	123 ÷ 12		
b	17 ÷ 5	g	17 ÷ 8	l	132 ÷ 13		
c	19 ÷ 3	h	23 ÷ 11	m	250 ÷ 6		
d	22 ÷ 7	i	57 ÷ 10	n	1456 ÷ 3		
e	31 ÷ 6	j	49 ÷ 7	o	2090 ÷ 18		

3 Find the batting average for each of the following boys and then put them in order, the best average first.

Innings	1st	2nd	3rd	4th	5th	6th	7th
A. Driver	7	3	6	6			
P. Willow	9	0	8	7	12		
C. Reese	0	9	11	9	15		
I. Field	15no	8	24	0	31	17	
S. Tumps	24	6	33	0	0	19	25no
P. Bales	23	27	0	54	42	37	11

4 Find the bowling average for each of the following boys and then put them in order, the best average first.

	wickets	runs		wickets	runs
P. Thrower	5	49	M. Pacer	7	129
S. Low	9	55	V. Green	15	177
A. Quick	11	67	A. Turner	21	151
T. Whirler	13	79	L. Mann	33	395

C Archery

Although bows and arrows have long been replaced as weapons of war, archery is today a popular and growing sport.

The British Standard Target has a diameter of **four feet** and there is a smaller target, used in international competitions, with a diameter of **80 cm**.

There is a gold circle in the centre of the target and this is ringed by four bands, red, black, and white in colour. The scores for each band are shown in the target below.

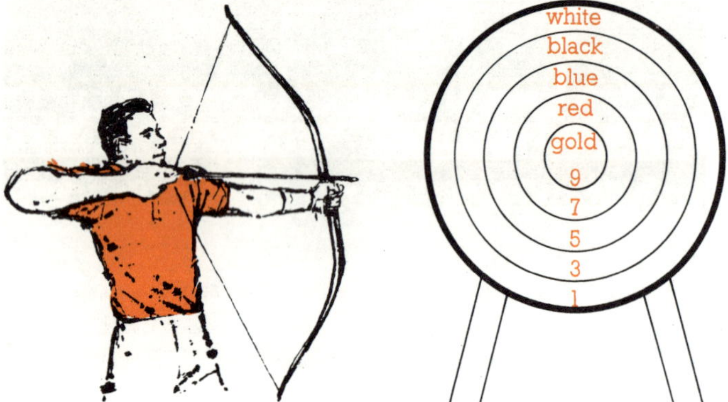

Usually each archer shoots six arrows and then the score is checked. In competitions an agreed number of arrows is shot from various distances.

1 Draw and colour a target twice the size of the one shown. Put in the score for each colour.

2 Calculate the total number of points scored on each of these targets. The crosses show where the arrows struck the target.

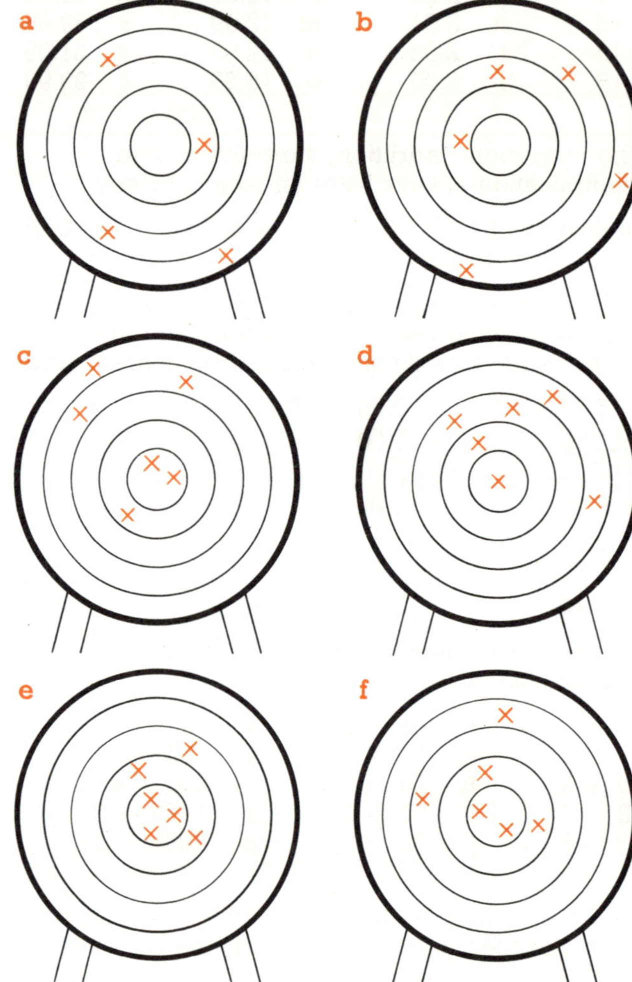

3 Calculate the total score for each of the following archers and put them in order, highest number of points first.

36 arrows at 30 metres

	gold	red	blue	black	white
A. Archer	2	10	12	6	3
S. Bowman	4	6	0	12	10
E. Shooter	3	4	9	11	9
P. Arrowsmith	5	7	6	10	4
I. Stringer	8	6	11	4	7

D Athletics

From the information below draw a graph to show the improvement in the times for the 100 m sprint in the Olympic Games.

Can you predict a time for the 100 m in the next Olympics?

1896	12.0s	1920	10.8s	1936	10.3s	1960	10.2s
1900	11.0s	1924	10.6s	1948	10.3s	1964	10.0s
1904	11.0s	1928	10.8s	1952	10.4s	1968	9.9s
1908	10.8s	1932	10.3s	1956	10.5s	1972	10.1s
1912	10.8s						

Set out your graph as shown.

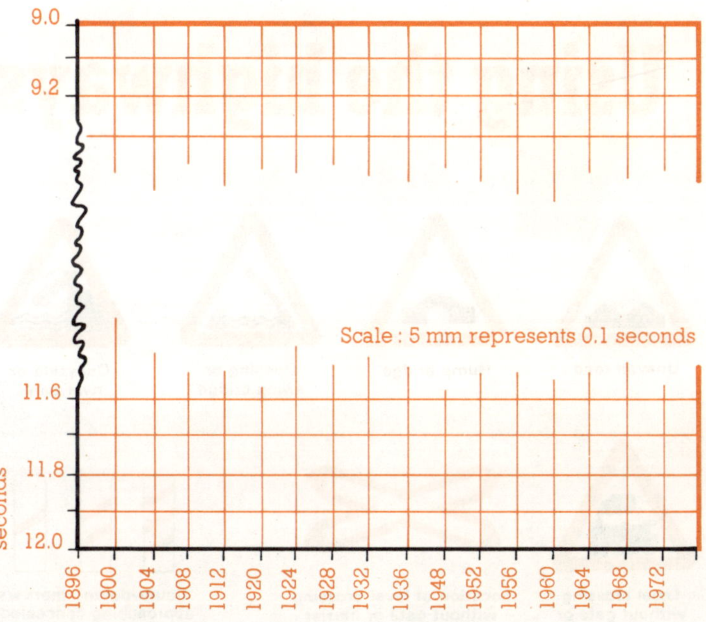

5

Using the highways

Whether we travel on foot or on wheels we all have to use the highways. Such things as traffic lights, zebra crossings, road signs and road markings have been provided to make the roads safer. We must all make sure that we understand them and know how to use them properly.

Uneven road

Hump bridge

Opening or swing bridge

Quayside or river bank

Level crossing without gate or barrier ahead

Location of level crossing without gate or barrier

"Count-down" markers approaching concealed level crossing

Assignments

A See how well you know the **Highway Code** by copying and completing these sentences.

1 If there is no pavement you should walk on the ▓▓▓▓▓▓▓ hand side of the road.

2 A group of people marching on the road should keep to the ▓▓▓▓▓▓▓

3 If the 'green man' on a road sign is flashing you should ▓▓▓▓▓▓▓ the road.

4 The important items to check on a bicycle are ▩▩▩▩▩▩▩

5 Double white lines along the middle of the road mean that ▩▩▩▩▩▩

6 A motorist must not sound his horn at night between ▩▩▩▩▩ and ▩▩▩▩▩

7 A ▩▩▩▩▩ traffic sign gives an order, a ▩▩▩▩▩ traffic sign gives a warning, and a ▩▩▩▩▩ traffic sign gives information.

8 The letters A.A. stand for ▩▩▩▩▩ and the letters R.A.C. stand for ▩▩▩▩▩

B Can you remember the order of the traffic-light signals? When you are certain copy these four sets of lights and then colour the circles to show the different lights that are lit up. Start off with red only.

C The bar chart shows the approximate shortest stopping distances for cars travelling at certain speeds. Study the chart carefully.

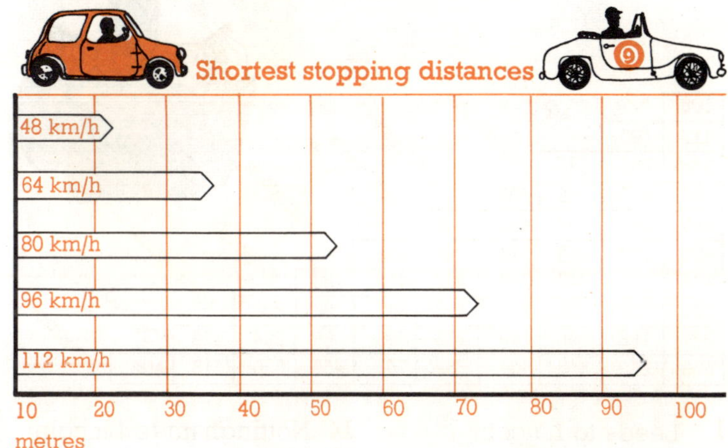

Now copy this table and fill in the missing numbers by referring to the bar chart.

Speed in km/h	48		64		80		96		112
Stopping distances in metres	23		37						
Difference		14							

To find the numbers for the **differences** line take each stopping distance from the following stopping distance. For example: 37 − 23 = 14 gives the first number on the **differences** line.

D Here is part of a distance chart. Find out how it works and then give the following distances. All distances are in kilometres.

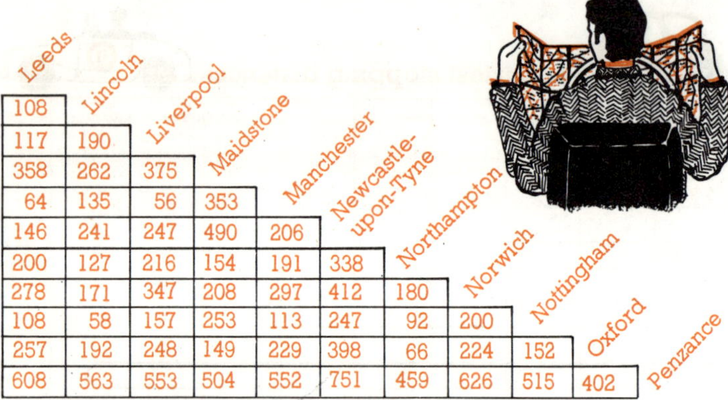

	Leeds	Lincoln	Liverpool	Maidstone	Manchester	Newcastle-upon-Tyne	Northampton	Norwich	Nottingham	Oxford
Lincoln	108									
Liverpool	117	190								
Maidstone	358	262	375							
Manchester	64	135	56	353						
Newcastle-upon-Tyne	146	241	247	490	206					
Northampton	200	127	216	154	191	338				
Norwich	278	171	347	208	297	412	180			
Nottingham	108	58	157	253	113	247	92	200		
Oxford	257	192	248	149	229	398	66	224	152	
Penzance	608	563	553	504	552	751	459	626	515	402

1 Leeds to Lincoln
2 Leeds to Liverpool
3 Lincoln to Liverpool
4 Lincoln to Manchester
5 Lincoln to Nottingham
6 Manchester to Maidstone
7 Maidstone to Leeds
8 Northampton to Nottingham
9 Liverpool to Nottingham
10 Nottingham to Newcastle upon Tyne
11 Newcastle upon Tyne to Lincoln
12 Norwich to Leeds
13 Norwich to Northampton
14 Nottingham to Lincoln
15 Maidstone to Norwich
16 Leeds to Newcastle upon Tyne
17 Norwich to Liverpool
18 Nottingham to Manchester
19 Manchester to Northampton
20 Maidstone to Manchester
21 Northampton to Lincoln
22 Lincoln to Norwich
23 Norwich to Newcastle upon Tyne
24 Manchester to Newcastle upon Tyne
25 Northampton to Leeds

E Practise the following constructions and then draw accurate scale drawings of the road signs shown in the margin. You will need a ruler and a pair of compasses.

To construct an equilateral triangle

A ———————————————— B
Draw a line **AB** 5 cm long

Set your compasses at 5 cm with **A** as centre draw an arc

Keep your compasses set at 5 cm and with **B** as centre draw a second arc to cut the first arc at **C**

Join **A** to **C** and **B** to **C**

To bisect an angle

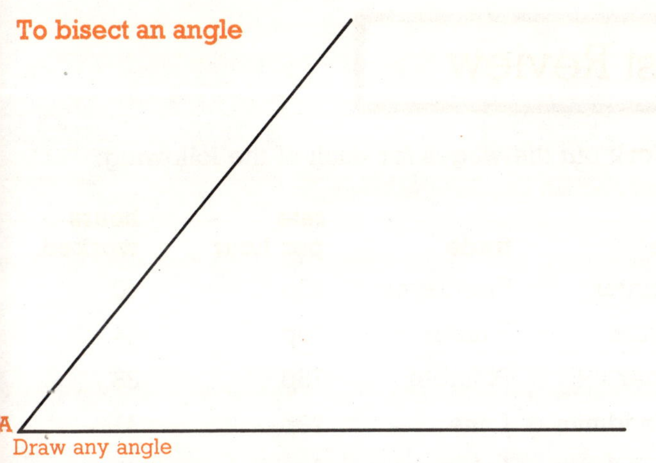

Draw any angle

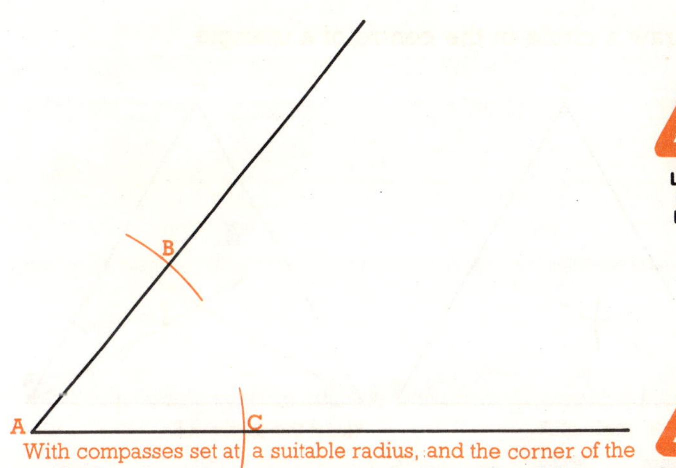

With compasses set at a suitable radius, and the corner of the angle **A** as centre, draw arcs to cut one arm at **B** and the other arm at **C**

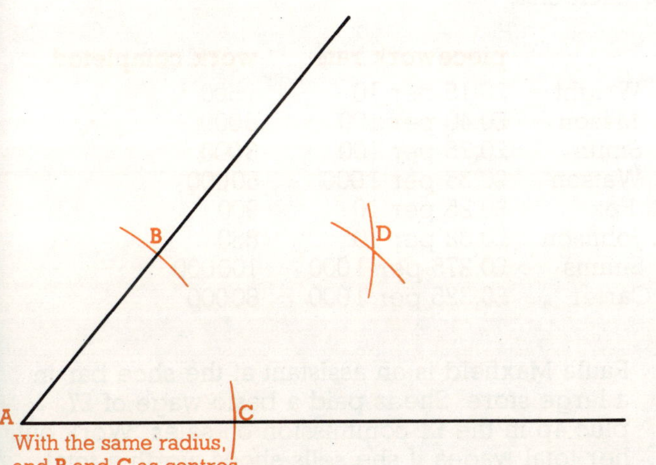

With the same radius, and **B** and **C** as centres, draw two arcs to cut at **D**

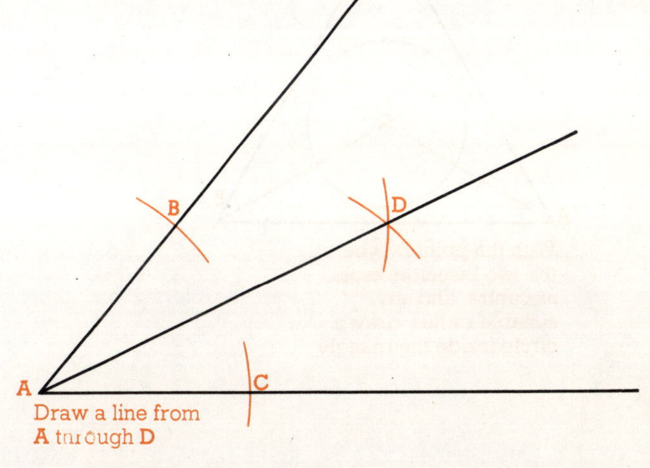

Draw a line from **A** through **D**

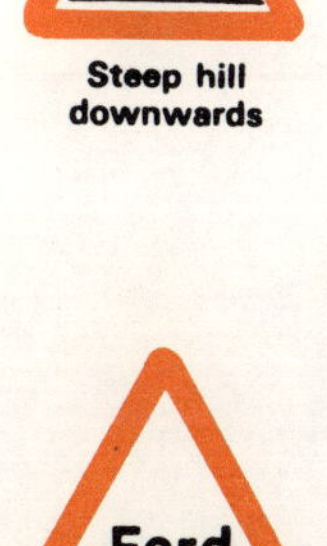

25

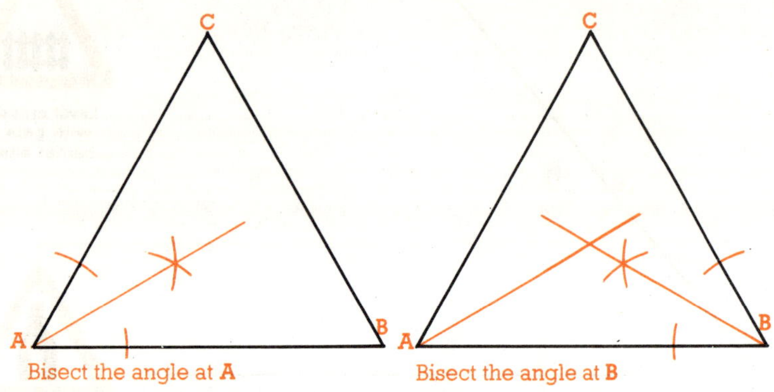

Bisect the angle at **A**

Bisect the angle at **B**

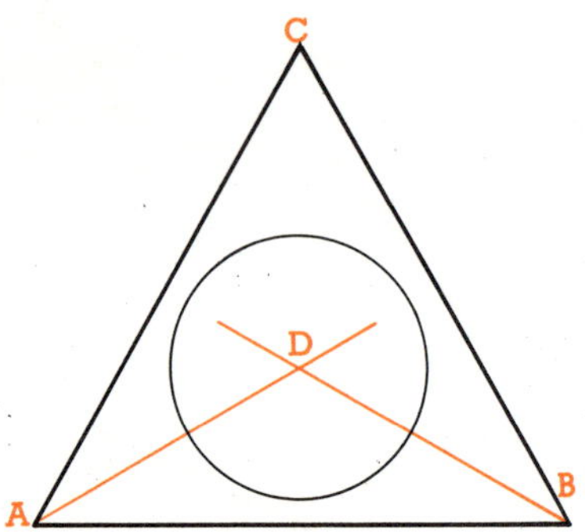

With the point **D**, where the two bisectors cross, as centre, and any suitable radius, draw a circle inside the triangle

1st Review

A Work out the wages for each of the following:

name	trade	rate per hour	hours worked
A. Pointer	Bricklayer	85p	40
B. White	Painter	70p	44
L. Piper	Plumber	75p	38
R. Woodman	Joiner	72p	41
I. Musselman	Labourer	60p	45

B The following girls are employed on a piecework basis in a clothing factory. Work out the wages for each one.

	name	piecework rate	work completed
1	L. Wright	£0.15 per 10	1 500
2	A. Mason	£0.40 per 100	5 000
3	T. Smith	£0.75 per 100	3 000
4	L. Watson	£0.35 per 1 000	60 000
5	M. Fox	£0.25 per 10	900
6	H. Johnson	£0.32 per 10	850
7	L. Simms	£0.275 per 1 000	100 000
8	I. Carter	£0.325 per 1 000	80 000

C 1 Paula Maxfield is an assistant at the shoe bar in a large store. She is paid a basic wage of £7, plus 4p in the £1 commission on sales. Work out her total wages if she sells shoes worth a total of £350 in a week.

2 Mr. Dawson is a commercial traveller selling breakfast cereals. He receives a basic wage of £15 plus 5% commission on sales. How much does he earn in a week during which his sales total £500?

3 Andrew Emson is an assistant in a tailor's shop. His weekly commission is 5% on sales up to £100 and 8% on sales over £100. His basic wage is £10 a week. How much does he earn in a week in which he sells clothes to a value of £420?

4 Mr. Hunt collects rents for a number of private landlords. He is paid $2\frac{1}{2}$% commission on all the rents he collects. How much does he earn if he collects £1 200 in one week?

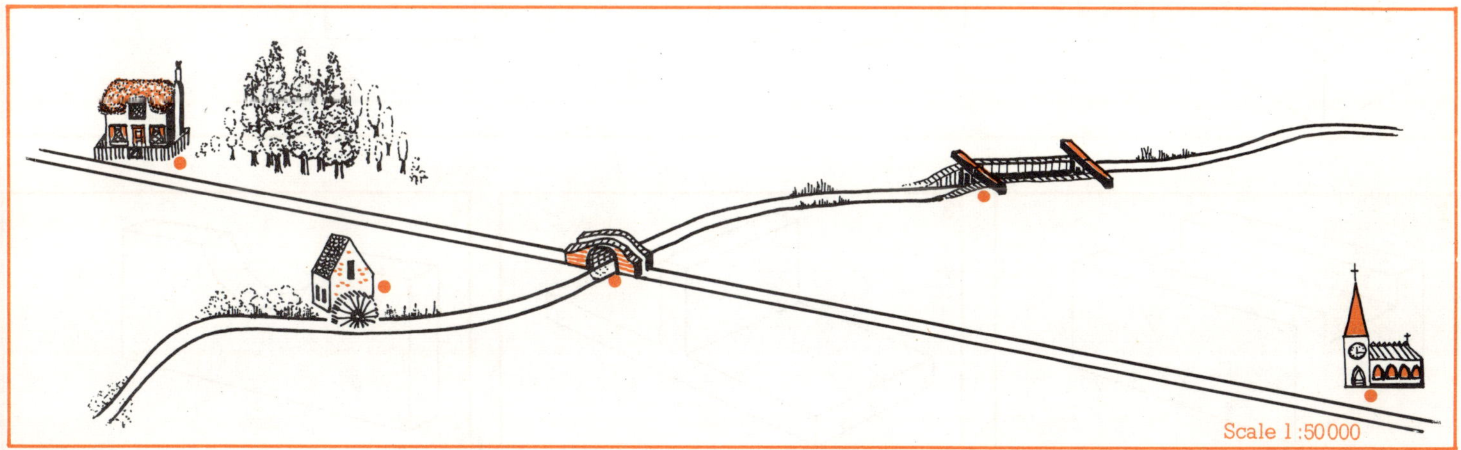

Cottage

Mill

Bridge

Lock gates

Church

D By measuring and using the **R.F.** find the approximate distances between:

1 the cottage and the bridge

2 the bridge and the church

3 the mill and the bridge

4 the church and the lock gates

5 the lock gates and the mill

6 the church and the mill

7 the cottage and the church

8 the bridge and the lock gates

 Copy the following sketches and draw a plan view and end elevation for each one. The first one has been done for you.

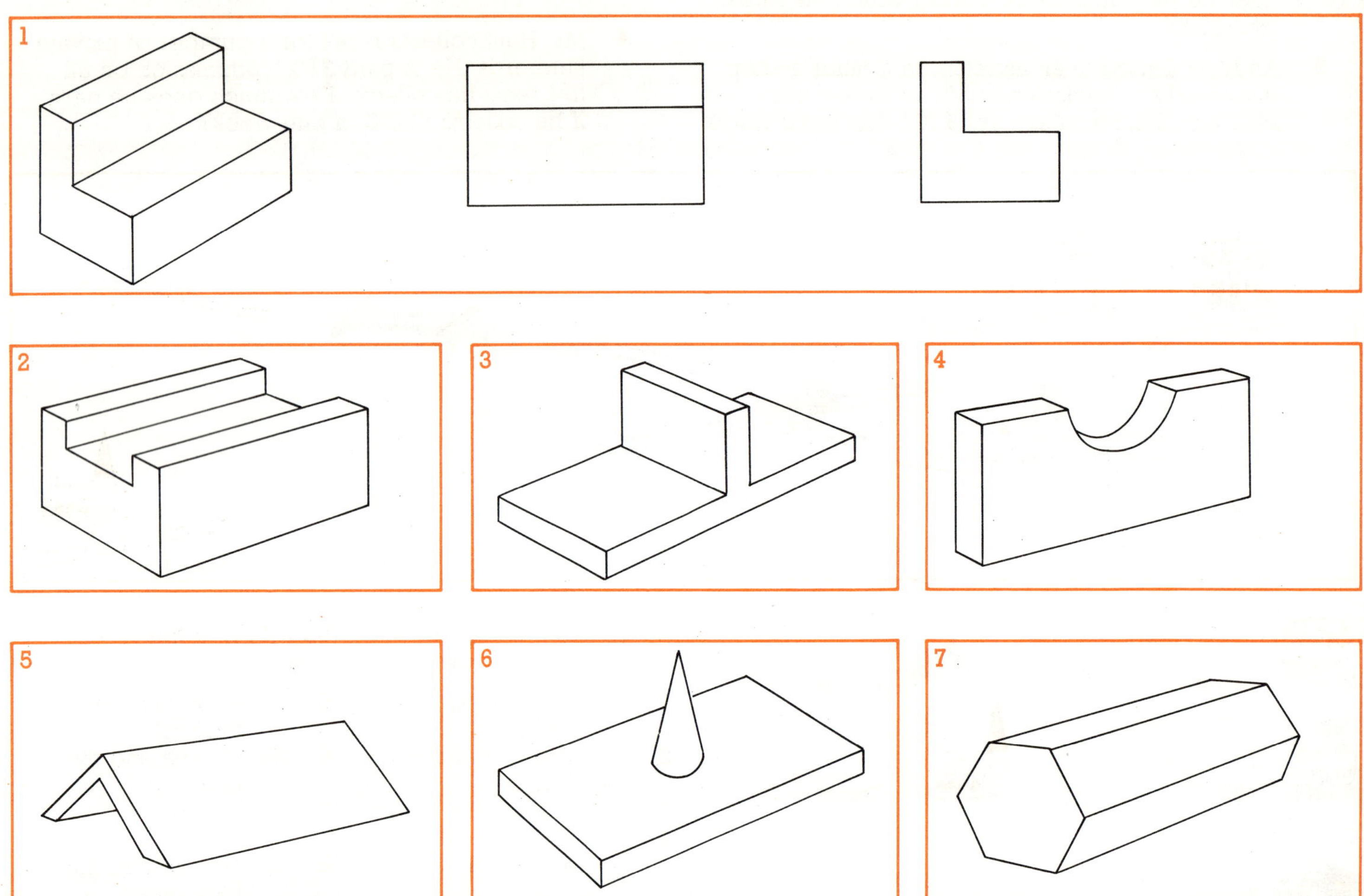

F Look at the postage rates for Inland Letters on page 12 and find the cost of sending the following by letter post:

1 Twenty-five letters, each weighing 30g, by first-class post

2 Ten packages, each weighing 200g, by first-class post

3 Five catalogues, each weighing 400g, by second-class post

4 Two hundred circulars, each weighing 20g, by second-class post

5 Six packages weighing 650g each, by first-class post

G Reduce the following message to the least possible number of words and then find the cost of sending your message by ordinary telegram.

> 'TO ALAN BROWN, SPORTS OUTFITTERS, 16 SOUTHSEA ROAD, LITTLEBOURNE. THE CONSIGNMENT OF CRICKET BATS HAS BEEN SENT TODAY AND THE CARRIERS HAVE PROMISED TO DELIVER THEM TO YOU BY FRIDAY. JOHN SPENCER AND COMPANY, 7 CHELSEA ROAD, LEEDS.'

H 1 If six teams play each other twice in a league competition how many matches will be played altogether?

2 If 16 teams take part in a knockout competition how many matches will be played altogether?

3 Find Peter's batting average if his scores for eight innings were: 24, 8, 0, 17, 30, 32 not out, 23, and 17.

4 a 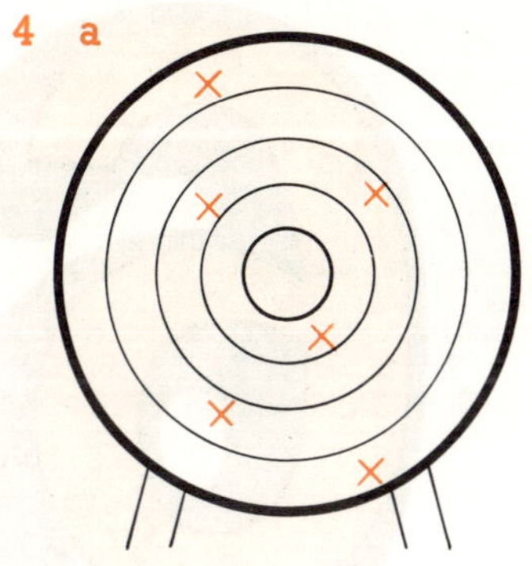

b c 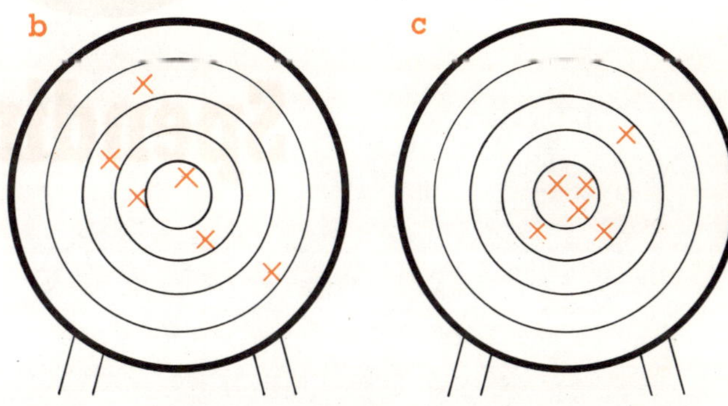

Calculate the total number of points scored on each of the above targets.

Most people want to earn high wages but it is just as important to spend your money wisely. In this chapter we are going to look at some of the ways of getting the best value for your money.

Planning a budget

Few people can buy everything they need and few people need everything they buy. If we are to spend, and save, our money sensibly then we have to write out a plan, that is a **budget**.

Here is an example. Barbara is 17, she works in an office and earns £12 a week. Her budget for one week is as follows:

6

Spending

At the end of each week Barbara puts her savings into the Post Office Savings Bank. Do you think this is a good way to spend her money?

Assignments

A 1   Copy out Barbara's budget shown on p. 30.

    2   Write out similar budgets for the following weeks. Remember that Barbara earns £12 a week and anything she has left she puts into savings.

Week ending January 14th

Board £3.50, Clothes £2.00, Cosmetics £0.35, Hairdresser £0.75, Travelling £0.60, Premium Bonds £1.00, Birthday present for friend £1.00, and Entertainment £0.50.

Week ending January 21st

Board £3.50, Clothes £1.20, Hairdresser £0.75, Travelling £0.60, Insurance £1.50, Premium Bonds £1.00, Entertainment £0.70, and Records £0.99.

Week ending January 28th

Board £3.50, Clothes £2.50, Cosmetics £0.49, Hairdresser £0.75, Travelling £0.60, Premium Bonds £1.00, and Hire Purchase deposit on a transistor radio £1.00.

Finding a bargain in the sales

It is often possible to save money by waiting until the shops have their January or Autumn sales.

To get the best bargains we have to find out which shops are offering the biggest reductions. To do this we must understand both decimals and percentages.

B   Remember:
    1   One percent (1%) means **one hundredth**.
    2   To find one hundredth of any quantity we divide by 100.
    3   To divide decimals by 100 we move the figures 2 places to the right.

Here are two examples

	£	£
Total amount (100%)	20.00	35.40
One percent of the total (1%)	0.20	0.354

Notice that £0.354 means thirty-five pennies and four-tenths of a penny. It is sometimes necessary to work in fractions of a penny.

Find 1% of the following amounts:

1	£10.00	11	£100.00	21	£1.00	31	£11.10
2	£20.00	12	£200.00	22	£2.00	32	£101.00
3	£30.00	13	£500.00	23	£5.00	33	£63.80
4	£60.00	14	£800.00	24	£9.00	34	£90.90
5	£12.00	15	£300.00	25	£3.00	35	£123.40
6	£16.00	16	£350.00	26	£3.50	36	£204.70
7	£19.00	17	£700.00	27	£4.00	37	£500.30
8	£11.00	18	£750.00	28	£4.50	38	£839.40
9	£40.00	19	£600.00	29	£7.00	39	£640.80
10	£70.00	20	£610.00	30	£7.20	40	£1234.50

C Here are three examples showing how to multiply sums of money.

```
      £              £                £
   0.12          3.16            23.57
×     7       ×     8        ×        4
   0.84         25.28            94.28
      1            1 4             12 2
```

Set out these examples as shown in the examples above.

1	£0.03 × 2	6	£0.12 × 3	11	£0.25 × 4
2	£0.04 × 3	7	£0.32 × 4	12	£1.23 × 5
3	£0.06 × 2	8	£0.52 × 5	13	£2.07 × 8
4	£0.08 × 5	9	£0.56 × 7	14	£3.82 × 9
5	£0.07 × 6	10	£0.81 × 9	15	£2.08 × 10

D We can now find a percentage of any sum of money.

Example

```
Find 24% of £12.50                              £
   1% of £12.50 = £0.125                      0.125
  24% of £12.50 = £0.125 × 24            ×       24
               = £3.00                        2.500
                                              0.500
                                              3.000
```

Calculate:

1	1% of £5.00	10	1% of £120
2	2% of £5.00	11	4% of £120
3	5% of £5.00	12	12% of £120
4	1% of £8.00	13	1% of £1.50
5	3% of £8.00	14	2% of £1.50
6	6% of £8.00	15	8% of £1.50
7	1% of £24.00	16	1% of £4.50
8	6% of £24.00	17	4% of £4.50
9	10% of £24.00	18	16% of £4.50

E Work out the sale prices for each item and find which store is offering the better bargains. The usual price is the same in each store.

WELLWORTHS

15% reduction offered on all instruments

SUPERMART

ALL instruments REDUCED by 10p in the £1 whilst the sale lasts

GUITARS
USUAL PRICE £15

BUY NOW and save 25% of the price of any record player

SPECIAL OFFER!

30p in the £1 off all Record Players bought during the sale

RECORD PLAYERS
USUAL PRICE £60

40p in the £1 off all records on this counter

ALL RECORDS half price while the sale lasts

PIANO TIME POP

POPULAR L.P.s.
Usual Price £2·50

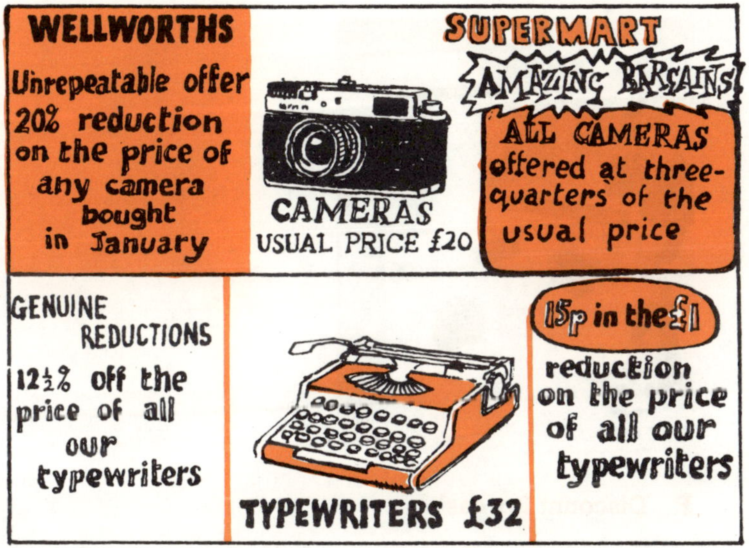

WELLWORTHS

Unrepeatable offer 20% reduction on the price of any camera bought in January

CAMERAS
USUAL PRICE £20

SUPERMART

AMAZING BARGAINS

ALL CAMERAS offered at three-quarters of the usual price

GENUINE REDUCTIONS

12½% off the price of all our typewriters

15p in the £1 reduction on the price of all our typewriters

TYPEWRITERS £32

F Discount for cash

If you buy an expensive article and pay for it immediately most shops are prepared to make a reduction in the price. This reduction is called **discount**.

Copy the following advertisement and price list and fill in the spaces.

ELECTRA AND COMPANY LTD

5% discount for cash allowed on all goods

	advertised price	discount	cash paid
Automatic Washer	£100	£5	£95
Refrigerator	£40	£2	
Food Mixer	£30		
Dish Washer	£200		
Spin Dryer	£20		
Cleaner	£45		
Health Lamp	£12		
Electric Razor	£15		
Hair Dryer	£13		
Electric Clock	£18		
Television	£250		
Electric Fire	£42		

7

Using electricity

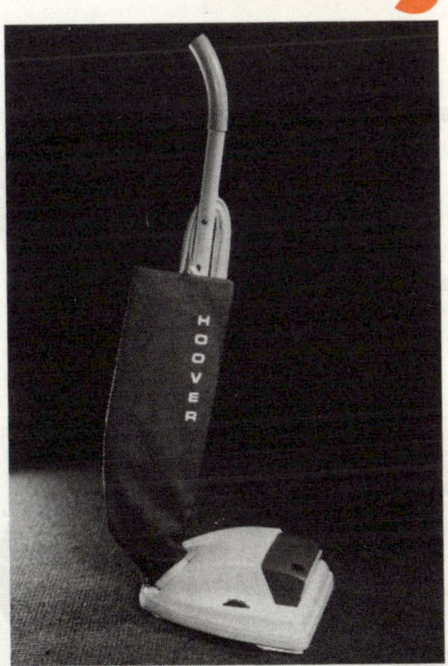

How many electric appliances have you at home? They can all be turned on at the touch of a switch but do you think of the cost of the electricity they are using?

In our own interests we should be able to read a meter, we should know how much electricity an appliance uses, and we should know about the different ways of paying for electricity.

Assignments

Reading a meter

The electricity we use first passes through a meter and the total amount is shown on a set of dials.

Here are the important things to know when you are reading a meter:

* There are six dials which are numbered alternately clockwise and anticlockwise. Can you think why?
* Each dial has a number above it which gives the value of the divisions on the dial.
* The $\frac{1}{10}$ dial is not taken into account when the meter is read.
* When the pointer is between two numbers we read the smaller number. The only exception is when the pointer lies between 9 and 0 in which case we read 9.

A Express the following numbers in words.

	10000	1000	100	10	1
1			1	0	0
2		1	0	0	0
3	1	0	0	0	0
4		2	0	0	0
5		2	5	0	0
6		2	0	5	0
7		2	0	0	5
8	6	2	0	0	0
9	6	0	0	2	0
10	6	0	2	0	0

	10000	1000	100	10	1
11	5	8	7	0	0
12	5	0	8	7	0
13	4	6	2	9	0
14	3	0	0	7	3
15	1	2	0	1	2
16	7	4	9	3	8
17	2	6	8	5	7
18	7	0	3	0	9
19	5	6	4	1	8
20	8	9	8	9	8

B Look carefully at this example and then give the readings on the dials alongside.

10000 units 2000 units 400 units 40 units 8 units

The metre reading is 12 448 units

1

2

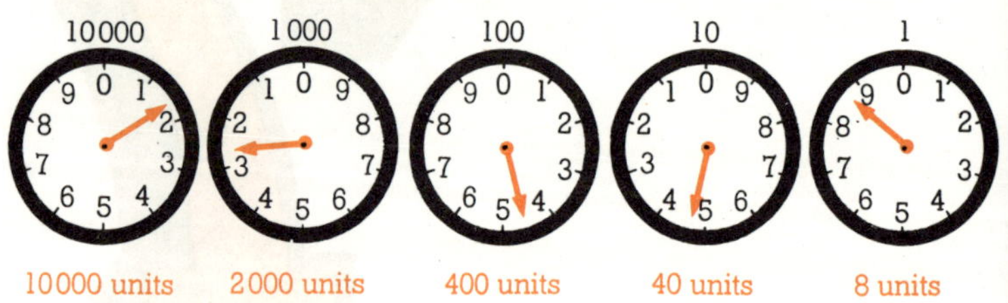

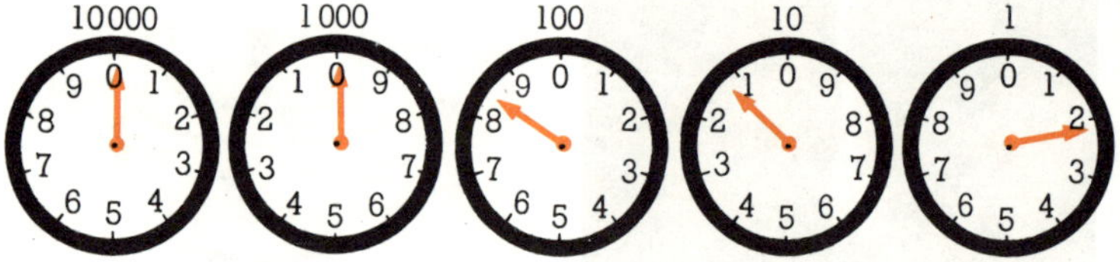

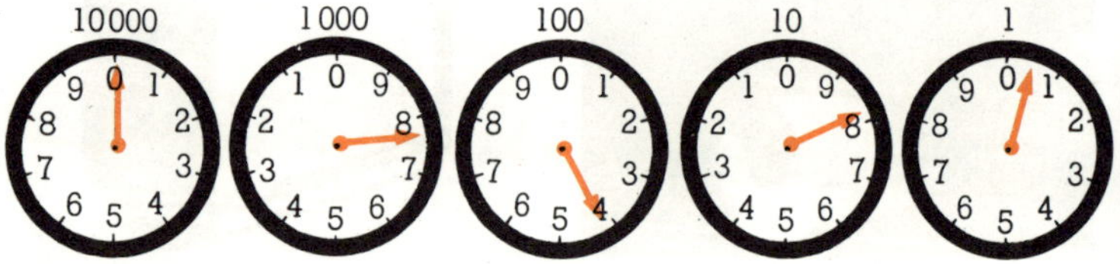

3
10000 1000 100 10 1
7
10000 1000 100 10 1
4
10000 1000 100 10 1
8
10000 1000 100 10 1
5
10000 1000 100 10 1
9
10000 1000 100 10 1
6
10000 1000 100 10 1
10
10000 1000 100 10 1

Paying the bill

C The cost of electricity varies from place to place and there are different rates for homes and factories.

Here is an example of an electricity bill. Notice that the Electricity Boards use decimals of a penny.

MR K WATTS	8 245 2540 24
22 BRIGHTSIDE LANE	2 FEB 1974 meter reading
LEEDS	7 FEB 1974 date of issue

meter readings			units charged at each rate			
present	previous	total units	3.728p	1.386p	0.808p	£
58371	57804	567	72	—	495	6.68
					total now due	6.68

Copy the following table and work out the numbers for columns **C** and **F**. If you have a calculating machine work out the **total due** to go in column **G**.

A	B	C	D	E	F	G
		Total	Units charged at each rate			Total due
Present	Previous	units	3.728p	1.386p	0.808p	£
47896	47344		72	—		
48399	47896		72	—		
48775	48399		72	—		
49206	48775		72	—		
49786	49206		72	—		
50311	49786		72	—		
50759	50311		72	—		
51302	50759		72	—		
51800	51302		72	—		
52387	51800		72	—		

Using electricity

Hair dryer
loading : 500 watts

Fan heater
loading : 2 kilowatt and 1 kilowatt

D The letters kWh on a meter stand for **kilowatt hour** and this is the unit measured by the meter.

A kilowatt hour is the amount of heat or energy produced by 1 kilowatt (1 000 watts) of electricity in 1 hour.

To find the cost of using an appliance we need to know:

a   the rate at which an appliance uses electricity (the loading).
b   the length of time the appliance is in use.
c   the cost of a unit of electricity.

$$\text{cost} = \frac{\text{loading in kilowatts}}{} \times \frac{\text{time in hours}}{} \times \frac{\text{cost per unit}}{}$$

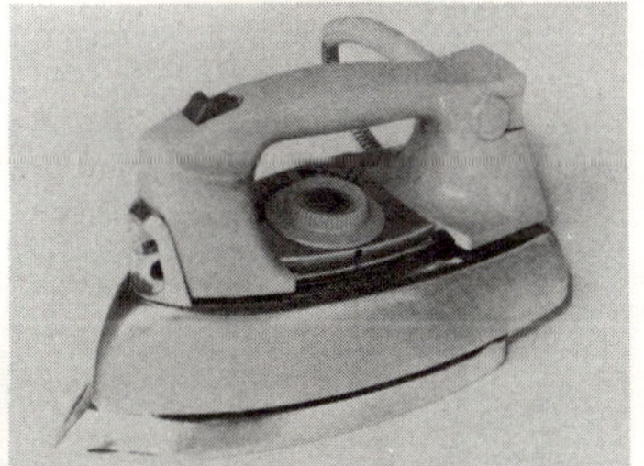

Here is an example

Find the cost of using a 500 watt iron for half an hour at 0.8p a unit.

The cost is: $(0.5 \times 0.5 \times 0.8)p = 0.2p$

watts W	100	200	250	400	500	750	1 000
kilowatts kW	0.1	0.2	0.25	0.4	0.5	0.75	1
minutes	10	15	20	30	45	50	60
hours	0.17	0.25	0.33	0.5	0.75	0.83	1

Find the cost of electricity for these examples.
You will find the above conversions helpful.

	Appliance	Loading	Time used	Cost per unit	Cost of electricity used
1	iron	1 000 W	1 hour	0.8p	
2	iron	1 000 W	2 hours	0.8p	
3	cooker oven	2 kW	1 hour	0.8p	
4	cooker oven	2 kW	$\frac{1}{2}$ hour	0.8p	
5	hair dryer	500 W	1 hour	0.8p	
6	hair dryer	500 W	$\frac{1}{4}$ hour	0.8p	
7	fan heater	3 kW	2 hours	0.8p	
8	fan heater	3 kW	15 min	0.8p	
9	razor	20 W	10 min	0.8p	
10	kettle	2.5 kW	10 min	0.8p	
11	toaster	1250 W	15 min	0.8p	
12	food mixer	150 W	20 min	0.8p	
13	fire	2 kW	45 min	0.8p	
14	cooker grill	2.5 kW	15 min	0.8p	
15	immersion heater	3 kW	$1\frac{1}{4}$ hrs.	0.8p	

8

Youth hostelling

If you enjoy the pleasures of the countryside then
practically all you need are the items pictured above.

The Youth Hostels Association (YHA) has provided
hostels for people walking, cycling and canoeing;
as the handbook says, '. . . for people travelling under
their own steam'.

Assignments

A Here are some of the things to do in the countryside, listed by the YHA. Write down these activities and then add at least six more you have thought of yourselves.

bird watching   cycling   orienteering

canoeing   mountaineering   photography

You will need these items for the assignments that follow: an atlas, an A.A. or R.A.C. handbook, a railway timetable, and a YHA handbook.

B The Lake District is one of the favourite areas for hostelling in this country. These assignments cover some of the preparations you would have to make for a hostelling holiday in the Lakes.

1 On a map of England, Scotland and Wales mark in your own town (or nearest town) and the Lakeland town of Keswick.

Use a printed map if possible, otherwise trace a map from an atlas.

2 **a** Find the direct distance from your own town to Keswick.

b Use an A.A. or R.A.C. handbook to find the distance by road from your own town to Keswick.

c Find the difference between the two distances.

3 Find out:

a the cost of the bus fare from your nearest bus station to Keswick.

b a list of convenient departure and arrival times for a bus journey to Keswick.

c the time the journey to Keswick takes on a bus.

The Lake District

C **1** Use the above map to find the walking distances between the following Lakeland hostels. Give each answer to the nearest kilometre.

Keswick to Buttermere   Ambleside to Patterdale

Buttermere to Eskdale   Patterdale to Longthwaite

Eskdale to Ambleside   Longthwaite to Keswick

Overnight fees	Young (over 5 and under 16)	Junior (16 and under 21)	Senior (21 and over)
Simple	29p	35p	41p
Standard	36p	44p	50p
Superior	44p	52p	62p
Special	51p	64p	80p

Meals

Supper 38p   Breakfast 29p   lunch packet 13p

2   Use your answers to question 1 to find the time it would take to walk from one hostel to the other at an average speed of 3 km/h.

3   Work out your own route for a tour of the Lake District and find the approximate distances between the hostels.

D   Since Y.H.A. hostels vary in the comfort and facilities they provide they are divided into four main grades : Simple, Standard, Superior, Special.

Normally hostellers are not allowed to stay more than three nights at a hostel.

The YHA supply booking forms similar to those shown below. Find the total charge in each case. The first example has been done for you.

1 HOSTEL BOOKING LETTER   To the Warden   LONGTHWAITE (SUPERIOR)   Youth Hostel

Kindly reserve the following—

Beds (Males) (See handbook for overnight fees)	☐ Young	1 Junior	☐ Senior
Beds (Females) (See handbook for overnight fees)	☐ Young	☐ Junior	☐ Senior
Meals please provide in respect of each night booked	1 Supper(s)	1 Break-fast(s)	☐ Lunch packet(s)

For the night of   THURSday   8TH AUGUST month   1974

	£
1 Junior male	0.52
1 supper	0.38
1 breakfast	0.29
total charge	1.19

2 HOSTEL BOOKING LETTER   To the Warden   KESWICK (STANDARD)   Youth Hostel

Kindly reserve the following—

Beds (Males) (See handbook for overnight fees)	☐ Young	1 Junior	1 Senior
Beds (Females) (See handbook for overnight fees)	☐ Young	☐ Junior	☐ Senior
Meals please provide in respect of each night booked	2 Supper(s)	2 Break-fast(s)	2 Lunch packet(s)

For the night of   FRI day   16TH August month   1974

3 HOSTEL BOOKING LETTER   To the Warden   BLACK SAIL (SIMPLE)   Youth Hostel

Kindly reserve the following—

Beds (Males) (See handbook for overnight fees)	☐ Young	☐ Junior	☐ Senior
Beds (Females) (See handbook for overnight fees)	☐ Young	3 Junior	☐ Senior
Meals please provide in respect of each night booked	3 Supper(s)	3 Break-fast(s)	3 Lunch packets

For the night of   WEDNES day   3RD JULY   month   1974

HOSTEL BOOKING LETTER

To the Warden **AMBLESIDE (SPECIAL)** Youth Hostel

Kindly reserve the following—

Beds (Males)
(See handbook
for overnight fees)
[ ] Young [ ] Junior [1] Senior

Beds (Females)
(See handbook
for overnight fees)
[ ] Young [2] Junior [1] Senior

Meals
please provide in
respect of each
night booked
[4] Supper(s) [4] Break-fast(s) [ ] Lunch packet(s)

For the night of _THURS_ day _5TH SEPT_ month _197 4_

HOSTEL BOOKING LETTER

To the Warden **PATTERDALE (SPECIAL)** Youth Hostel

Kindly reserve the following—

Beds (Males)
(See handbook
for overnight fees)
[ ] Young [1] Junior [1] Senior

Beds (Females)
(See handbook
for overnight fees)
[ ] Young [1] Junior [1] Senior

Meals
please provide in
respect of each
night booked
[4] Supper(s) [4] Break-fast(s) [4] Lunch packet(s)

For the night of _THURS_ day _18TH JULY_ month _197 4_

HOSTEL BOOKING LETTER

To the Warden **BUTTERMERE (SUPERIOR)** Youth Hostel

Kindly reserve the following—

Beds (Males)
(See handbook
for overnight fees)
[ ] Young [1] Junior [1] Senior

Beds (Females)
(See handbook
for overnight fees)
[ ] Young [1] Junior [ ] Senior.

Meals
please provide in
respect of each
night booked
[3] Supper(s) [ ] Break fast(s) [ ] Lunch packets

For the night of _TUES_ day _5TH AUGUST_ month _197 4_

9

Algebra

Algebra is an important branch of mathematics and it is used by people in many different occupations.

Not only does algebra help us to write statements in a short and simple way, it also gives us quick and easier methods of working things out.

The first assignments will help you remember how to write in the 'language' of algebra. The remaining assignments give practice in using algebra.

Assignments

A Write these statements in a shorter form.

Example

$$a + 3b + 2a - 2b = 3a + b$$

1. $a + a$
2. $a + 2a$
3. $a + 2a + 3a$
4. $2b + 5b$
5. $3b + b + 4b$
6. $a + a + b + b$
7. $3a + 2b + a + 3b$
8. $m + 2n + 5n + 3m$
9. $2x + 7y + 8x + 3y$
10. $2b + 3b + 4b + 5a$
11. $4c + 3d + a + 2c$
12. $x + 2x + y + 3y$
13. $5p + 2q - 2p - q$
14. $10a + 3b + 7b - 7a$
15. $5s + 6t - 5s + 6t$
16. $a + ab + a + ab$
17. $ab + 2bc + 5ab + 3bc$
18. $4xy + yz - xy - yz$
19. $5pq + 2qr - 3pq + 2qr$
20. $6abc + a + 2abc$
21. $ab + ba$
22. $3ba + 2ab$
23. $5xy - 5yx$
24. $3abc + 2bac$
25. $9ab + 2ba - 10ab$
26. $abc + acb + bca$
27. $2xyz + 5xzy$
28. $4pq + 4pq + 4qp$
29. $3abc + 4abc - 7cba$
30. $2pqr + 3qrp + 4rpq$

B Write these statements in a shorter form.

Example

$$(b \times 2 \times c) + (b \times d \times 3) = 2bc + 3bd$$

1 3×4
2 4×3
3 $a \times b$
4 $b \times a$
5 $3 \times a$
6 $a \times 4$
7 $5 \times a \times b$
8 $c \times 2 \times d$
9 $e \times f \times 7$
10 $3 \times a \times b \times c$
11 $e \times 1 \times f$
12 $2 \times a \times 3 \times b$
13 $5 \times c \times d \times 2$
14 $ab \times c$
15 $a \times bc$

16 $3 \times ab \times c$
17 $x \times 2y \times z$
18 $2a \times 3b$
19 $4p \times 5q$
20 $p \times 2q \times 3r$
21 $(a \times b) + (c \times d)$
22 $(5 \times c \times d) + (6 \times e)$
23 $(a \times 7 \times b) + (a \times 2 \times b)$
24 $(3 \times a) - (2 \times a)$
25 $(q \times 5) - (3 \times q)$
26 $cd + (2 \times c \times d)$
27 $3ef + (4 \times e \times f)$
28 $(a \times b \times c) + 3abc$
29 $(3p \times q) + (2 \times pq)$
30 $(p \times 0) + (0 \times q)$

C Write these statements in fraction form.

Example

$$3a \div 4b = \frac{3a}{4b}$$

1 $p \div q$
2 $q \div p$
3 $4 \div 5$
4 $5 \div 4$
5 $a \div 3$

6 $3 \div a$
7 $2a \div 3b$
8 $2b \div 3a$
9 $x \div 5y$
10 $ab \div cd$

11 $5ac \div 6ef$
12 $7xy \div 8ab$
13 $9abc \div 10efg$
14 $10 \div 3fgh$
15 $(a + b) \div 2$

D The use of brackets

$3(a + b) \longrightarrow$ this means add a and b, then multiply the answer by 3.

$3a + 3b \longrightarrow$ this means multiply a by 3, multiply b by 3, then add the two answers.

The following diagram shows that $\boxed{3(a + b) = 3a + 3b}$

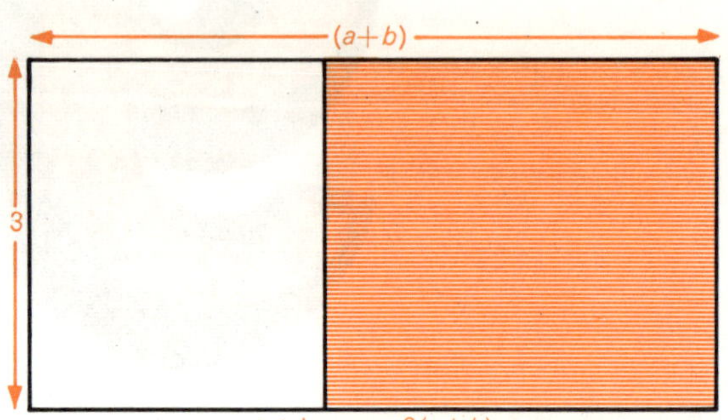

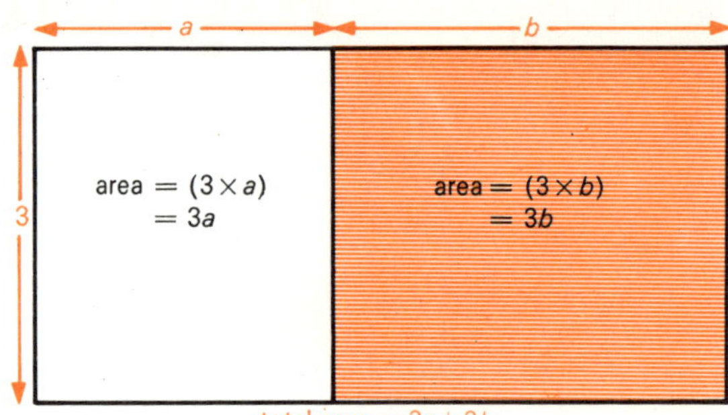

Here are four more examples:

$4(1+3) = (4\times 1)+(4\times 3)$
$4(c+3) = 4c+(4\times 3)$
$4(c+d) = 4c+4d$
$a(b+c) = ab+ac$

Write the following statements as the sum of two terms as shown in the above examples.

1 $2(a + b)$	11 $a(1 + b)$	21 $3a(c + d)$
2 $2(a + 1)$	12 $2a(1 + b)$	22 $3a(bc + de)$
3 $2(3 + 1)$	13 $c(b + 2)$	23 $4a(2b + 3c)$
4 $4(c + d)$	14 $3c(b + 2)$	24 $2(m + n)$
5 $4(2 + c)$	15 $f(2g + 3h)$	25 $\frac{1}{2}(m + n)$
6 $4(2 + 5)$	16 $2f(2g + 3h)$	26 $2(\frac{1}{2}m + n)$
7 $1(x + y)$	17 $3p(4q + 5r)$	27 $2ab(c + 2d)$
8 $x(1 + y)$	18 $5(ab + bc)$	28 $6xy(3 + 2z)$
9 $a(b + c)$	19 $a(1 + bc)$	29 $7p(3q + 2qr)$
10 $e(f + g)$	20 $a(bc + de)$	30 $3ab(5c + 1)$

E Formulae are rules written in numerals, letters and mathematical symbols. They are used by people in such varied occupations as farming, engineering, building and office work.

Here are some examples of useful formulae. Copy the first example which has been done for you in each case.

The train has a good average speed

The car travels fast on the motorway, slowly in traffic.

1 The average speed for a journey is given by the formula:

$$S = \frac{D}{T}$$

$S \longrightarrow$ average speed in km/h,
$D \longrightarrow$ distance in km,
$T \longrightarrow$ time in hours

$$S = \frac{D}{T}$$

$$= \frac{150}{3} \text{ km/h}$$

$$= 50 \text{ km/h}$$

Find the average speeds for the following journeys:

a 150 km in 3 hours   **b** 200 km in 4 hours
c 225 km in 5 hours   **d** 636 km in 6 hours

2 If a painter needs to work out the area of the four walls of a room he can use the formula:

$$A = 2h(l + b)$$

$A \longrightarrow$ area of the room,
$h \longrightarrow$ the height of the room,
$l \longrightarrow$ length of the room,
$b \longrightarrow$ the breadth of the room.

$$A = 2h(l + b)$$
$$= 2 \times 2.5(4 + 3)\text{m}^2$$
$$= 5 \times 7\text{m}^2$$
$$= 35 \text{ m}^2$$

Find the areas of the walls of these rooms.

	height	length	breadth
a	2.5 m	4 m	3 m
b	3 m	5 m	4 m
c	2.5 m	6 m	4 m
d	4 m	10 m	8 m

3   Fahrenheit and Celsius thermometers are both in common use. Here is a formula to change temperatures from the Fahrenheit to the Celsius scale:

$$C = \frac{5}{9}(F - 32)$$

Find the temperatures Celsius equivalent to the following temperatures Fahrenheit:

a   59° F   **b**   41° F   **c**   68° F   **d**   32° F

$$C = \frac{5}{9}(F - 32)$$
$$= \frac{5}{9}(59 - 32)°$$
$$= \frac{5}{9} \times 27°$$
$$= 15°$$

4   Builders sometimes use this formula when constructing the staircase in a house:

$$R = \tfrac{1}{2}(60 - T)$$

$$R \longrightarrow \text{rise} \qquad T \longrightarrow \text{tread}$$

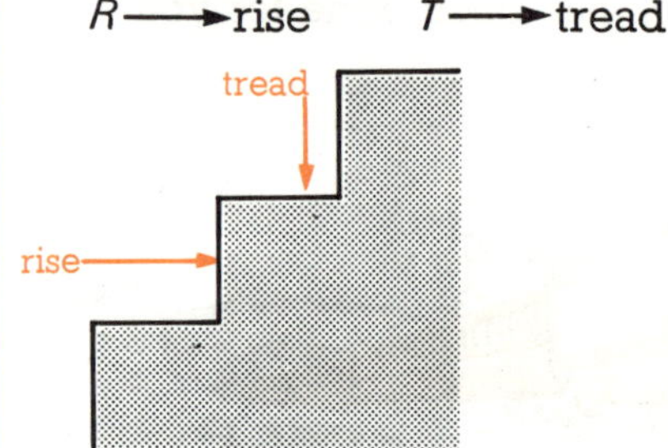

Work out the rise when the treads are:

a   26 cm   **b**   28 cm   **c**   30 cm   **d**   32 cm

$$R = \tfrac{1}{2}(60 - T)$$
$$= \tfrac{1}{2}(60 - 26)\text{ cm}$$
$$= \tfrac{1}{2} \times 34\text{ cm}$$
$$= 17\text{ cm}$$

5 Simple interest is worked out using the formula:

$$I = 0.01 (PRT)$$

$I \longrightarrow$ simple interest $P \longrightarrow$ the amount of money
$R \longrightarrow$ the rate of interest $T \longrightarrow$ the time in years

Find the simple interest for the following:

a £250 at 5% for 3 years **c** £300 at 6% for 4 years
b £100 at 4% for 2 years **d** £150 at 10% for 5 years

$$I = 0.01 \ (PRT)$$
$$= £ \ (0.01 \times 250 \times 5 \times 3)$$
$$= £ \ (2.50 \times 15)$$
$$= £37.50$$

10 Saving

Most of us need to save for things that cost a lot of money but what is the best way to save?

If we are sensible we shall want our money to be safe and we shall want to make our savings work by earning interest for us.

There are many organizations which offer to save people's money, we shall look at some of the more well-known ones but you will find it useful and interesting to find out about others for yourselves.

The National Savings Bank

This bank is run by the Government for ordinary people and it is part of the business of 21 000 post offices throughout the country.

A Copy these sentences and fill in the blanks by obtaining the information from a Post Office.

1 The National Savings Bank will accept deposits of ▆▆▆▆▆ p or more.

2 Deposits are not repayable for a child under ▆▆▆▆▆▆ years of age.

3 One withdrawal a day of up to £ ▆▆▆▆▆ can be made at any Savings Bank Post Office.

4 The present rate of interest on ordinary accounts is ▆▆▆▆ %

5 The first £ ▆▆▆▆ of interest is free of income tax.

B Below and on the next page you will see a typical page of a Savings Bank account book.

On demand	£ 10

This means that you have taken £10 out of the bank

Thirty pounds	£ 30

This means that you have paid £30 into the bank

balance	£ 45

This means that you have £45 left in the bank

Make a careful copy of the following account and notice how the balance is worked out.

Date of deposit	Amount of deposit or method of withdrawal	Th	H	T	U	Intls
	balance FIVE pounds	–	–	–	5 00	
July 6	On demand				4 00	RS
	balance				1 00	
Oct 22	Twenty five pounds			2 5	00	RS
	balance			2 6	00	
Oct 28	On demand			1 4	00	MG
	balance			1 2	00	
Dec 13	On demand			1 0	00	RS
	balance			2	00	
Jan 2	Thirty pounds			3 0	00	DM
	balance			3 2	00	

C Copy the following account and work out the balance after each entry. Put in your own dates and initials.

Date of deposit	Amount of deposit or method of withdrawal	£				Intls
		Th	H	T	U	
	balance TEN pounds		1	0	00	
	Fifteen pounds		1	5	00	
	balance					
	On demand			3	00	
	balance					
	On demand			5	00	
	balance					
	On demand			8	00	
	balance					
	Twelve pounds		1	2	00	
	balance					
	Eight pounds			8	00	
	balance					
	On demand			7	00	
	balance					
	On demand			9	00	
	balance					

Premium Savings Bonds

An unusual way of saving is provided by Premium Bonds which can be bought from any Post Office. No interest is paid on the Bonds but holders do have a regular chance of winning a prize.

D Copy the following sentences and fill in the blanks by obtaining the information from a Post Office.

1 Premium Bonds are sold in units of £⸺

2 The maximum number of Premium Bonds a person can hold is ⸺

3 Premium Bonds can be bought by anyone over ⸺ years of age.

4 There is a single weekly prize of £⸺

5 There are over 90 000 monthly prizes ranging from £⸺ to £⸺

6 All prizes are free of Income Tax.

1 How many Premium Bonds could you buy in a year if you saved

   a   £1 a week,   **b**   50p a week,   **c**   25p a week?

2 John bought 5 Premium Bonds on the first day of each month throughout a year. He won a £25 prize in May and a £50 prize in July and bought more Bonds with the money. How many Bonds did he own at the end of the year?

3 On each of his 16th, 17th and 18th birthdays Peter was given £8 in Premium Bonds. He won one prize of £100 and one prize of £25 and bought more bonds with the money. How many bonds did he then own?

4 Paul bought 10 Premium Bonds each month for two years and did not win any prizes.

   James bought 10 Premium Bonds each month for two years and won a £25 prize.

   Ian saved £10 a month in a bank for two years and received a total of £12.50 interest.

   How much money did each boy have at the end of two years?

5 About 90 000 prizes are paid to Premium Bond Holders each month. How many prizes are paid out in a year?

6 Imagine you have won a £25 Premium Bond Prize and explain what you would do with the money.

Building Societies

Many people prefer to save their money with a Building Society because they do not have to pay income tax on the interest that their money earns. Although the Societies do have their own special rules, in practice customers can deposit or withdraw money as they please, just as with an ordinary bank.

The longer a customer leaves his money in a Building Society the bigger the rate of interest he is offered. This is very useful for people who are saving up for expensive things such as a car or a house.

F   Overleaf is a Table showing how your money 'grows' if you save regularly with a Building Society.

Period of Saving	Monthly Savings					
	£1	£3	£5	£10	£20	£30
1 year	£12	£37	£62	£124	£248	£372
2 years	£25	£77	£128	£255	£510	£766
3 years	£39	£118	£197	£394	£789	£1 183
4 years	£54	£163	£271	£542	£1 084	£1 626
5 years	£70	£209	£349	£698	£1 397	£2 095
6 years	£86	£259	£432	£864	£1 728	£2 592
7 years	£104	£312	£520	£1 040	£2 080	£3 120
8 years	£122	£368	£613	£1 226	£2 452	£3 679
9 years	£142	£427	£711	£1 423	£2 847	£4 271
10 years	£163	£490	£816	£1 633	£3 266	£4 899

Copy the following examples and fill in the blanks by finding the information from the above table.

	Monthly savings	Period of saving	Savings plus interest
1	£3	5 years	£209
2	£1	2 years	
3	£1	4 years	
4	£5	10 years	
5	£10	5 years	
6	£3	8 years	
7	£20	6 years	
8	£10	7 years	
9	£30	4 years	
10	£5	8 years	
11	£1		£86
12		6 years	£432

G Use the table on this page to find the amount of interest paid in the following cases. Remember that this is compound interest at 6%.

Example

Find the amount of interest earned by saving £5 a month for four years.

Savings for 1 month	£5
Savings for 1 year	£5 × 12 = £60
Savings for 4 years	£60 × 4 = £240
Savings plus interest	£271 (from the table)
Amount of interest	£271 − £240 = £31

1 £1 a month saved for 3 years

2 £3 a month saved for 2 years

3 £5 a month saved for 2 years

4 £10 a month saved for 4 years

5 £10 a month saved for 8 years

6 £3 a month saved for 6 years

7 £5 a month saved for 10 years

8 £30 a month saved for 2 years

9 £20 a month saved for 5 years

10 £2 a month saved for 4 years

H Make the necessary enquiries and then copy and complete these sentences.

1 Building Society branches in our town include▓▓▓▓▓▓

2 The least amount needed to start a deposit account in the ▓▓▓▓▓ Building Society is▓▓▓▓▓ p.

3 The most that a person can invest in a Building Society is £▓▓▓▓▓

4 The present rates of interest paid by Building Societies are:

Deposit Accounts ▓▓▓▓▓▓ %.

Share Account ▓▓▓▓▓ %.

Subscription Shares ▓▓▓▓▓ %.

5 One of the main advantages of saving with a Building Society is that ▓▓▓▓▓ ▓▓▓▓▓ does not have to be paid on the interest.

2nd Review

A Here are details of John's finances for two weeks. Make out a budget account for each of the two weeks, entering anything he has left as savings.

Week ending July 10th

Wages £14.50, Board £4.00, clothes £1.50, travelling 60p, hairdresser 30p, cricket club fee £2.50 and entertainment £2.20.

Week ending July 17th

Wages £15.20, Board £4.00, clothes £1.20, deposit on a bicycle £3.00, cricket club outing £5.45 and insurance premium £1.10.

B Calculate:

1	1% of £10	**5**	5% of £25	**9**	12% of £15
2	3% of £10	**6**	2% of £32	**10**	7% of £50
3	8% of £10	**7**	10% of £120	**11**	4% of £2.50
4	1% of £25	**8**	6% of £32	**12**	25% of £85

C Find the number of 10 cm × 10 cm tiles that would be needed to cover the following areas.

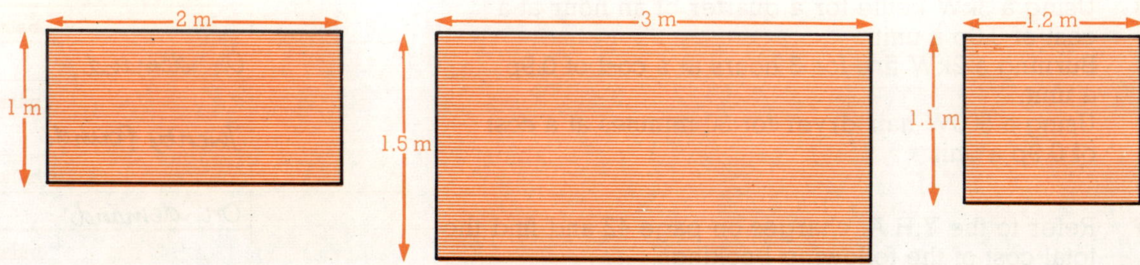

D Give the reading on each of the following sets of electric meter dials.

E Find the cost of electricity for each of these examples.

1 Using a 2kW cooker oven for half an hour at a cost of 0.8p a unit.
2 A 1kW fan heater switched on for one hour at a cost of 0.8p a unit.
3 Using a 3kW kettle for a quarter of an hour at a cost of 0.8p a unit.
4 Burning a 2kW fire for 3 hours at a cost of 0.9p a unit.
5 Using a 500W hair dryer for 30 minutes at a cost of 0.9p a unit.

F Refer to the Y.H.A. charges on page 42 and find the total cost of the following bookings.

1 Standard Hostel

Supper, bed and breakfast for two nights for three boys aged 17.

2 Superior Hostel

Supper, bed and breakfast for one night for mother, father, a boy aged ten and a boy aged fifteen.

3 Simple Hostel

Bed and breakfast for one night for four girls and one lunch packet for each.

G Copy this account and work out the balance after each entry. Put in your own date and initials.

Date of deposit	Amount of deposit or method of withdrawal	£				Intls
		Th	H	T	U	
	balance TWENTY pounds		2	0	00	
	Seven pounds			7	00	
	balance					
	Eleven pounds		1	1	00	
	balance					
	On demand			2	50	
	balance					
	On demand		1	2	00	
	balance					
	Twenty pounds		2	0	00	
	balance					
	On demand		1	7	50	
	balance					

H The formula $S = 90(2n - 4)$ gives the sum of the angles of a polygon.

S stands for the sum of the angles, n stands for the number of sides.

Here is the working for a five-sided figure, a pentagon.

$$S = 90(2n - 4)°$$
$$= 90(2 \times 5 - 4)°$$
$$= 90(10 - 4)°$$
$$= 90 \times 6°$$
$$= \mathbf{540°}$$

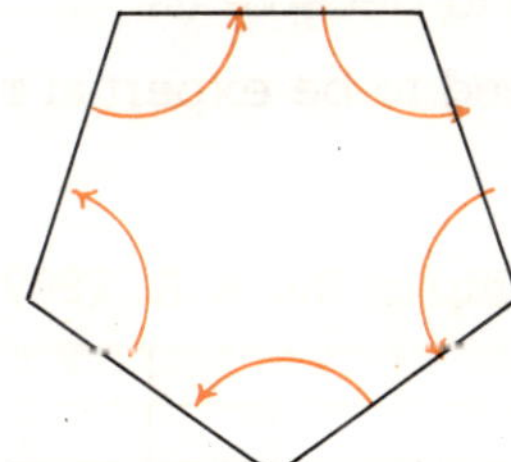

Use the formula to find the sum of the angles of these figures and then check your answers by measuring with a protractor.

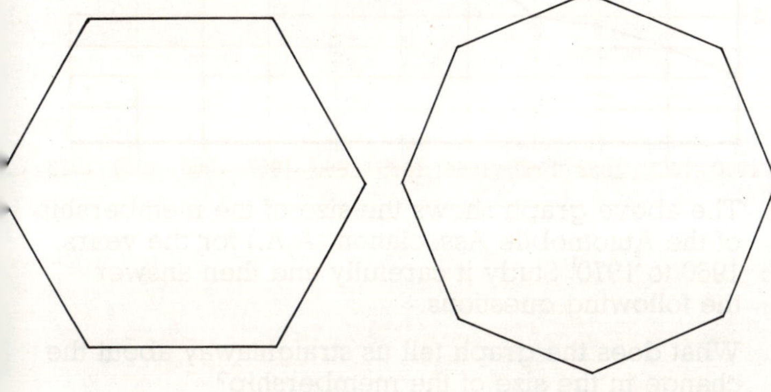

I This table shows how savings of £1 a month 'grow' over a period of 10 years if interest is paid at 6%

Year	1	2	3	4	5
Amount	£12	£25	£39	£54	£70
Year	6	7	8	9	10
Amount	£86	£104	£122	£142	£163

Draw a bar chart to show this information in the form of an advertisement for a Building Society.

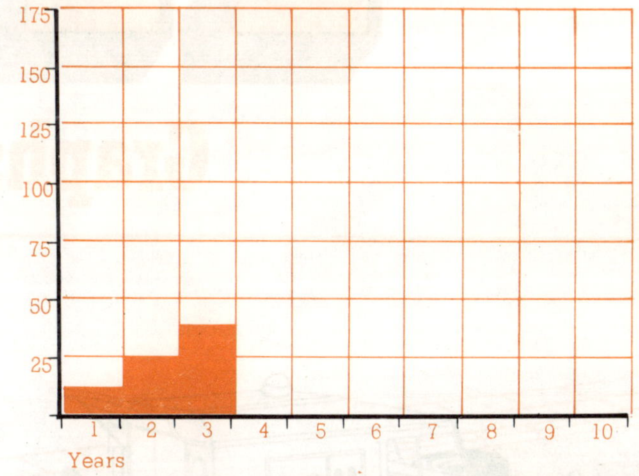

J Two brothers, Alan and Brian, were both left £150 by a rich aunt. Alan bought premium bonds with his money and won a £25 prize in the first year.

Brian put his money in a Building Society paying interest at the rate of 8%.

Which boy was better off after a year, and by how much?

11 Graphs

We have seen that there are several different kinds of graphs and that they can be used for a variety of purposes.

Graphs can be used to give information, to work out problems, or to keep records. They are used to persuade us to buy things and they are even used to mislead and deceive us.

We all need to be expert at reading and understanding graphs.

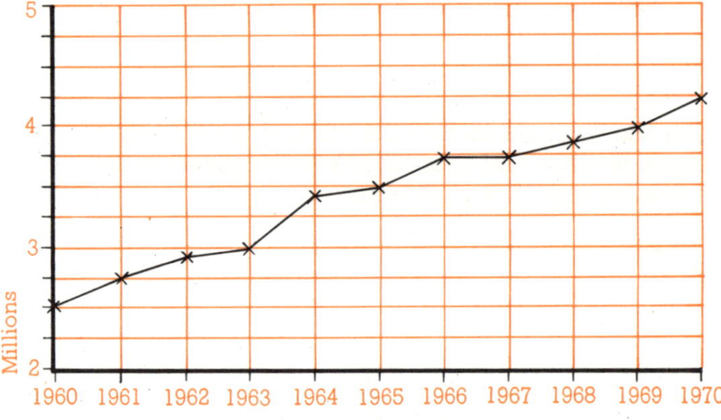

Membership of the A.A. 1969 to 1970

A The above graph shows the size of the membership of the Automobile Association (A.A.) for the years 1960 to 1970. Study it carefully and then answer the following questions.

1 What does the graph tell us straightaway about the change in the size of the membership?

2 What was the membership to the nearest quarter of a million
  a in 1960   **b** in 1970?

3 What was the increase in membership from 1960 to 1970?

4 What was the average yearly increase in membership from 1960 to 1970?

5 In which year was the increase greatest?

6 In which year was there no increase in membership?

Time and Distance

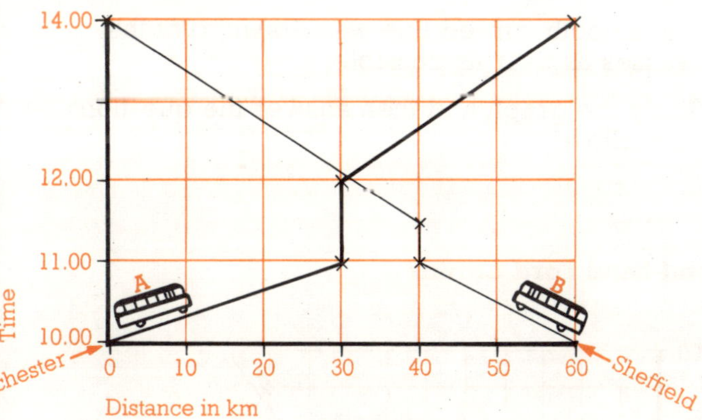

B Study the above graph very carefully and then answer the following questions.

1 How long does it take bus **A** to travel halfway to Sheffield?

2 What is the average speed of bus **A** over the first half of the journey?

3 What is happening to bus **A** between 11.00 and 12.00?

4 How long does it take bus **A** to complete the second half of the journey?

5 What is the average speed of bus **A** over the second half of the journey?

6 At what time does bus **A** arrive in Sheffield?

7 What is the average speed of bus **B** over the first 20 km of its journey?

8 For how long does bus **B** stop?

9 At what distance from Manchester do the buses pass each other?

10 How can you tell from the graph that the two buses have the same average speed for the whole journey?

11 What is the average speed for both buses over the whole journey?

12 What time is 14.00 hours in the 12-hour clock system?

C The above pie charts show what men and women think of their chances of avoiding serious injury if they are wearing a seatbelt.

Study the charts carefully and then copy the following sentences, choosing the correct words from those given in brackets.

1 (Men, Women) have more faith in seatbelts than (men, women).

2 About a (half, third, quarter) of both men and women think they have the same chance of avoiding injury whether they are wearing a seatbelt or not.

3 About (quarter, half, three-quarters) of the women think they have a better chance if they are wearing a seatbelt.

4 About a twelfth of the (men, women) think they have a worse chance with a seatbelt.

5 None of the (men, women) think they have a worse chance with a seatbelt.

6 Use your protractor to measure the angles at the centre of the charts and then work out the fractions for each of the following groups. One of the fractions has been worked out for you.

Men who think they have a worse chance

angle at the centre 30°

30° as a fraction of 360° $= \dfrac{30}{360} = \dfrac{1}{12}$

	Men		Women
Better		Better	
Same		Same	
Worse	$\dfrac{1}{12}$	Worse	

D Below are the fractions for the combined views of men and women about their chances of avoiding serious injury when wearing a seatbelt.

Draw a pie chart to show this information

Better $\dfrac{7}{10}$ **Same** $\dfrac{1}{4}$ **Worse** $\dfrac{1}{20}$

E The following graph was made up from the prices of second-hand cars advertised in a paper in May 1973.

Each cross represents the year and price of a Ford Cortina.

The smooth curved line was drawn to fit the crosses as near as possible.

Study the graph and then answer the questions that follow.

Second-hand Ford Cortina

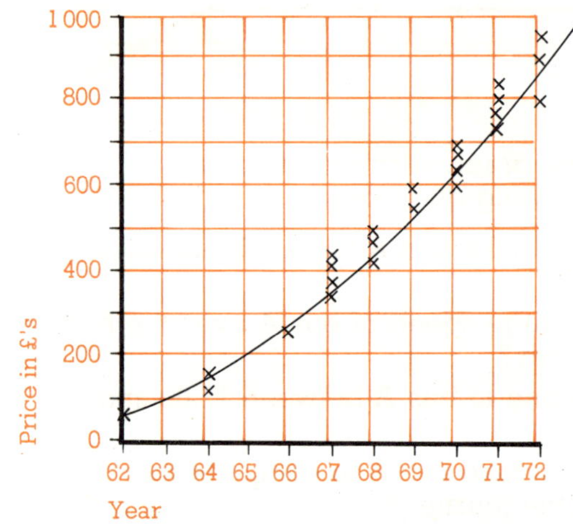

1 In which year does the difference in prices appear to be greatest?

2 In which year does the difference in prices appear to be smallest?

3 There were no advertisements for Cortinas registered in 1963 or 1965. What would have been the likely cost for Cortinas registered in those years?

4 What was the price for the 1962 Cortina?

5 What was the lowest price for a 1971 Cortina?

6 What was the highest price for a 1971 Cortina?

7 What would have been a possible price for a 1961 Cortina?

F

1 Obtain a local paper giving the prices of second-hand cars. Make a note of the prices of Cortinas registered over the past 10 years. Draw a graph similar to the one on the previous page.

2 Arrange for the class to draw similar graphs for other makes of cars.

G How Savings Grow

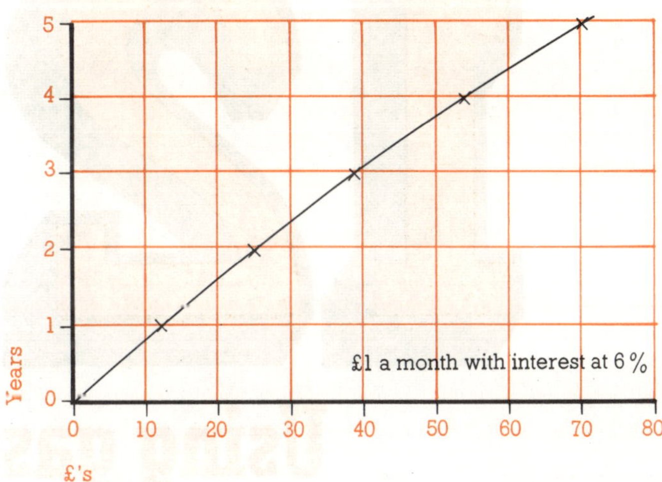

Copy and complete the following table using the graph for the second line.

Amount of savings

	Number of years				
	1	2	3	4	5
£1 a month saved at home	£12	£24			
£1 a month saved in a Building Society paying 6% interest	£12	£25			

12 Using gas

For well over a hundred years the gas used in our homes and factories was produced from coal but in 1965 the first major strike of natural gas was made in the North Sea. Since then four large gas fields have been found and a large part of the country has been converted to natural gas.

A Reading a meter

When gas enters a house from the mains it first passes through a meter which measures the volume of gas which is used. There are several different kinds of meters but most are similar to the one shown below.

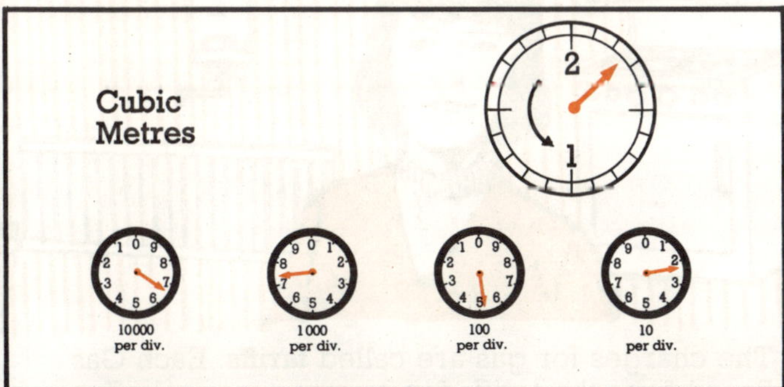

Gas meters are read in the same way as electric meters but there is no dial for units. The reading from the above meter is

67 520

Make a copy of the following dials and give the reading for each one.

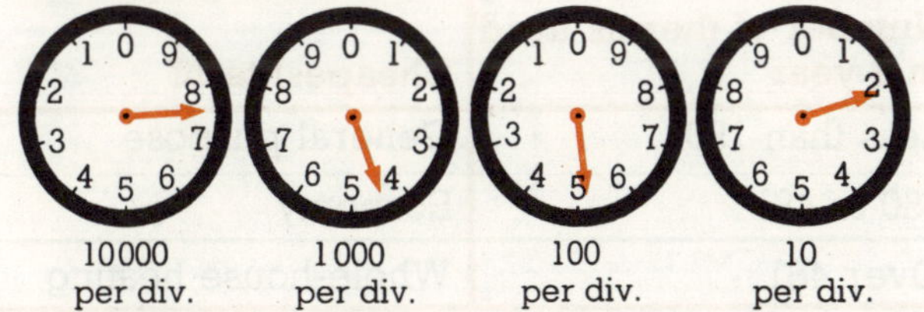

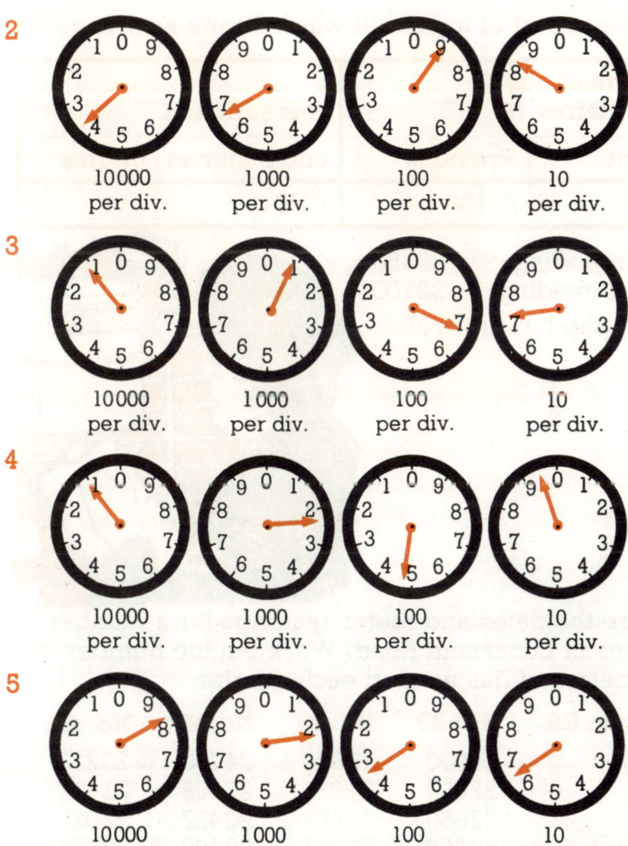

B The Gas Bill

Usually the 'gas man' calls every three months to read the meter. The Gas Board computer then works out the amount of gas you have used and the amount you have to pay, and you are then sent a bill.

Here is part of a gas bill which shows how it works.

Meter readings cubic metres		Gas supplied	
Present	Previous	cubic metres	therms
23 450	22 100	1 350	

Present reading	23 450
Previous reading	22 100
Gas supplied in m³	1 350

Here are the dates and meter readings for a number of houses in Laburnum Road. Work out the number of cubic metres of gas used in each quarter.

Laburnum Rd.	No. 23	No. 25	No. 27	No. 29
January 4	21 000	19 820	24 763	20 223
April 7	22 400	21 030	26 008	21 612
July 3	22 950	21 465	26 427	22 020
October 4	23 586	22 077	26 989	22 539
January 5	24 699	22 989	27 899	23 444

Laburnum Rd.	No. 31	No. 33	No. 35	No. 37
January 4	19 375	17 112	18 786	20 101
April 7	20 469	18 537	19 998	21 237
July 3	21 001	19 061	20 537	21 716
October 4	21 367	19 480	20 888	22 143
January 5	22 108	20 377	21 003	23 429

C At one time the old gas companies used to charge for the **volume** of gas they supplied. Since the heat obtained from gas varies, however, the Gas Boards now charge for the amount of **heat** supplied.

The amount of heat produced by gas is measured in **therms** and you will see this heading on the gas bill shown in Assignment B.

The charges for gas are called **tariffs**. Each Gas Board fixes the tariffs for its own area and offers a choice of tariff to its customers. The cheapest tariff to use depends on the number of therms used in a year: if you choose the wrong tariff you pay more for your gas than you need.

Here is an example from one Gas Board that supplies natural gas.

Number of therms used in a year	Cheapest tariff
Less than 150	General purpose
150 to 481	Economy
Over 481	Whole-house heating

Write down the following headings:

General purpose   Economy   **Whole–house heating**

and then put the following numbers under the tariff which will provide the cheapest gas.

Therms used in a year		Therms used in a year		Therms used in a year	
1	100	9	260	17	407
2	200	10	149	18	307
3	300	11	151	19	88
4	400	12	395	20	129
5	500	13	495	21	316
6	600	14	700	22	478
7	145	15	134	23	487
8	155	16	164	24	138

D Customers who have a gas cooker and gas central heating are likely to burn over 1000 therms a year. Their cheapest way of paying for gas will be to use the Whole-house Heating tariff.

Here is an example of a Whole-House Heating tariff.

Quarterly standing charge	£4.00
Price per therm	6p

The quarterly charge tells us that customers receive a gas bill every three months.

For the following examples:

a work out the cost of gas for each quarter on the Whole-House Heating tariff,

b find the total cost of gas for the year.

The first and second quarters of the first example have been done for you.

First quarter	£
440 therms at 6p a therm	26.40
Quarterly standing charge	4.00
amount due	30.40

Second quarter	£
250 therms at 6p a therm	15.00
Quarterly standing charge	4.00
amount due	19.00

	First quarter	Second quarter	Third quarter	Fourth quarter
1	440 therms	250 therms	100 therms	320 therms
2	400 therms	200 therms	120 therms	300 therms
3	420 therms	230 therms	80 therms	340 therms
4	510 therms	220 therms	110 therms	350 therms
5	426 therms	195 therms	124 therms	305 therms

A sink water heater burns 0.4 therms an hour.

Find:

a   the number of therms burned
b   the cost of the gas at 6p a therm

if the sink heater is in use for:

1   2 hours   3   5 hours   5   12 hours
2   3 hours   4   10 hours   6   24 hours

A gas fire burns 0.12 therms an hour.

Find:

a   the number of therms burned
b   the cost of the gas at 6p a therm

if the gas fire is in use for:

7   2 hours   9   10 hours   11   24 hours
8   5 hours   10   20 hours   12   100 hours

13 Mathematical designs

Everywhere around us we see mathematical designs. On television, in our homes, on our clothes, and even sometimes on our food.

Many of these designs are based on a circle and in this chapter we are going to look at a number of examples.

Copy the designs in your book, using the given measurements, and then draw one of them on a larger scale and colour it in.

You can then make up some examples of your own.

A A crown

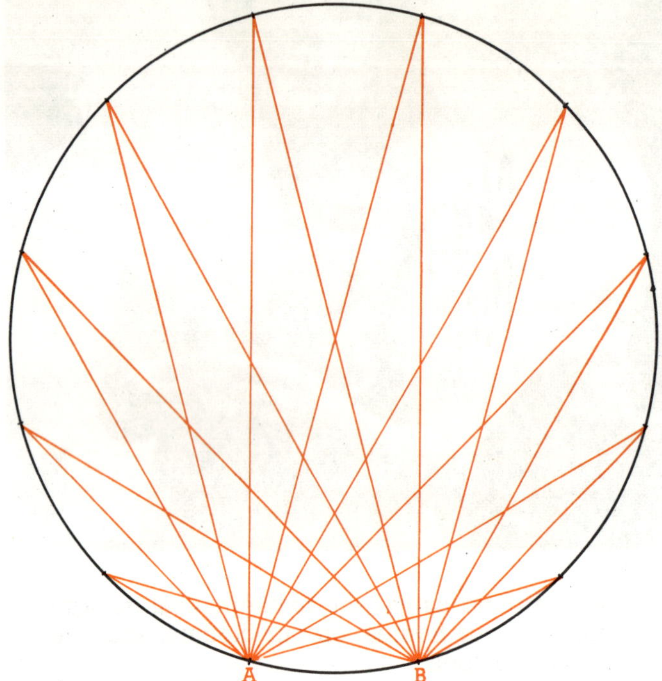

B A star

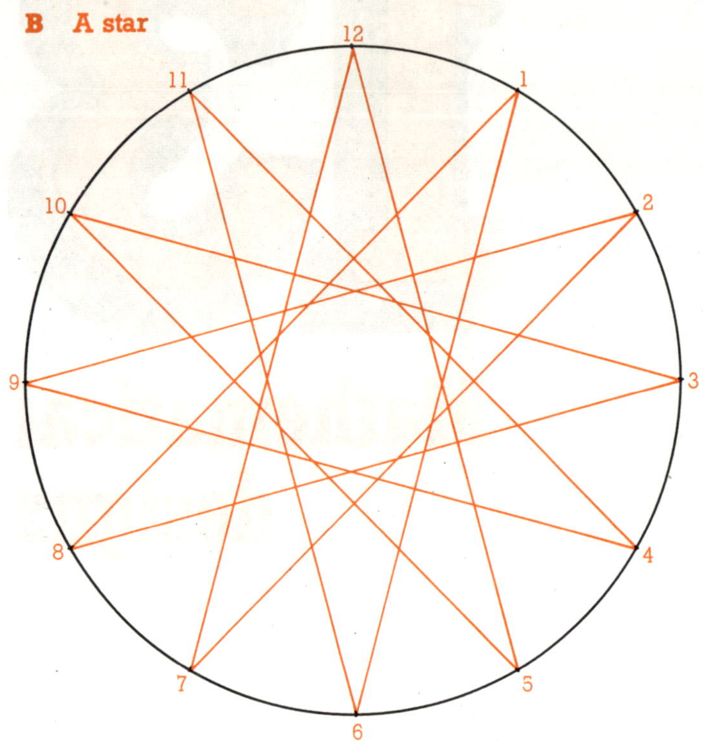

1 Draw a circle radius 6 cm.
2 Mark off 12 points at equal distances round the circumference.
3 Choose two adjacent points **A** and **B**.
4 Join **A** and **B** to each of the other points round the circle, but not to each other.

1 Draw a circle radius 6 cm.
2 Mark off 12 points at equal distances round the circumference
3 Number the points 1 to 12.
4 Copy the above design.

C A diamond

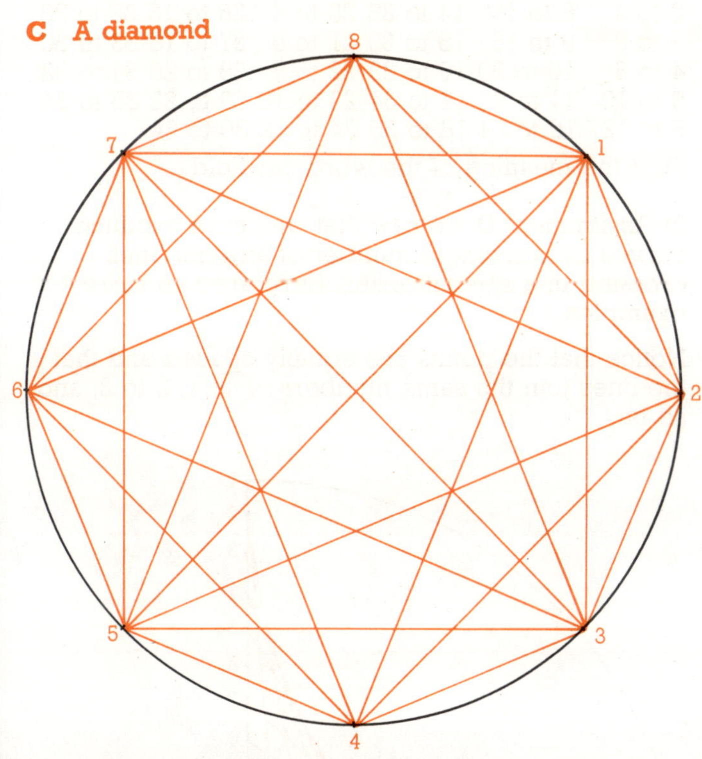

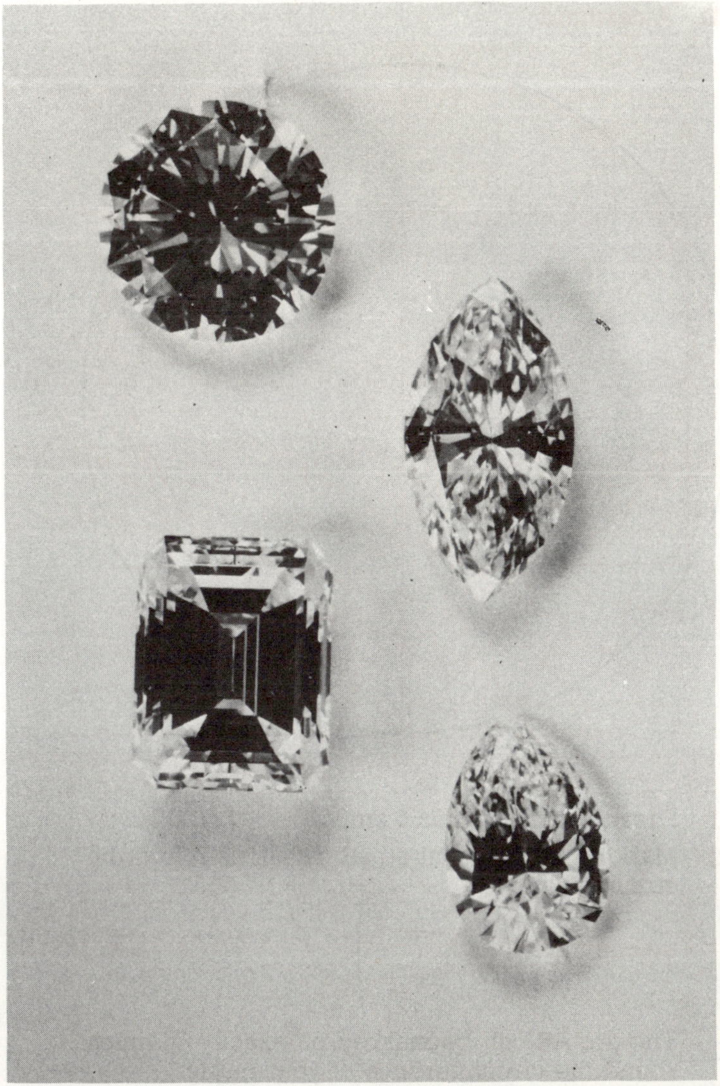

1 Draw a circle radius 6 cm.

2 Mark off 8 points at equal distances round the circumference.

3 Number the points 1 to 8.

4 Copy the above design.

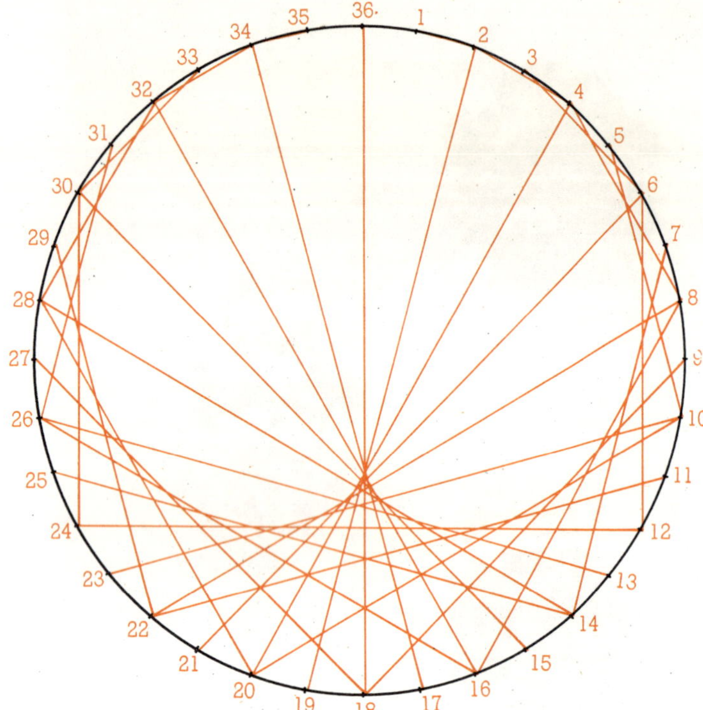

1 Draw a circle radius 6 cm.

2 Mark off 36 points at equal distances round the circumference.

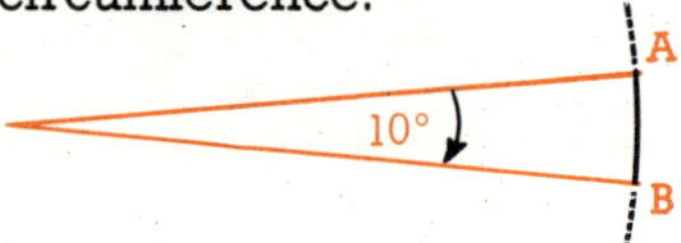

The arc **AB** can be marked off exactly 36 times round the circumference of your circle.

3 Number the points 1 to 36.

4 Join the following points:

1 to 2	7 to 14	13 to 26	19 to 2	25 to 14	31 to 26
2 to 4	8 to 16	14 to 28	20 to 4	26 to 16	32 to 28
3 to 6	9 to 18	15 to 30	21 to 6	27 to 18	33 to 30
4 to 8	10 to 20	16 to 32	22 to 8	28 to 20	34 to 32
5 to 10	11 to 22	17 to 34	23 to 10	29 to 22	35 to 34
6 to 12	12 to 24	18 to 36	24 to 12	30 to 24	

5 Find the meaning of the word cardioid.

E In Assignment **D** we saw that we could produce curves by drawing a number of straight lines crossing in a special order. Here are two more examples.

Notice that the points are equally spaced and that the lines join the same numbers; 1 to 1, 2 to 2, and so on.

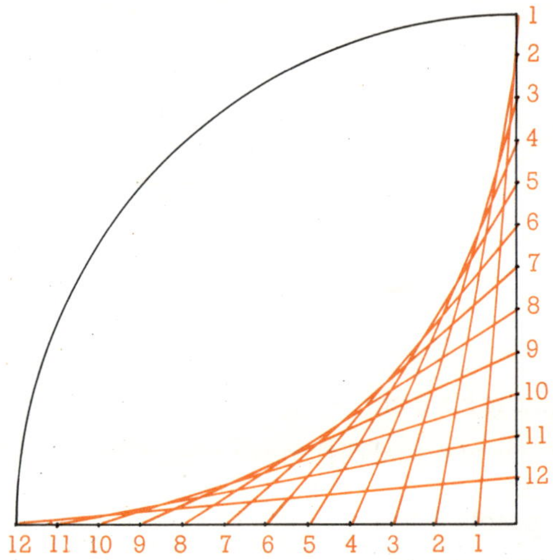

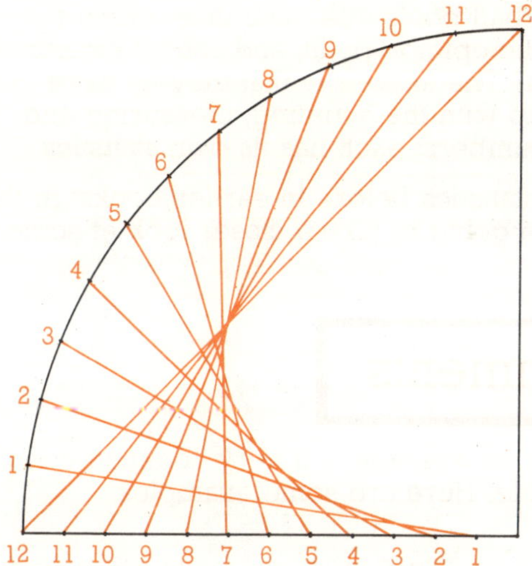

Remove the paper from the board and hammer a
nail in each of the points.

You can then interweave the thread to trace the
straight lines on your drawing, and so reproduce
your design.

We can use this idea to produce a design for our
own home or classroom.

The equipment required is:
a piece of drawing paper,
a piece of blockboard painted black,
some panel pins or small nails,
a hammer and some white thread.

When you have drawn a satisfactory design on the
drawing paper, position the paper on the blockboard
with drawing pins, and then mark the required
points on the board, using a hammer and a nail.

14

Statistics

What have football pools, life insurance, cricket averages, public opinion polls, and school examinations got in common? The answer is that they all have something to do with the counting, measuring and recording of numbers: each has its own **statistics**.

We have met statistics before in earlier books: in this chapter we are going to take a closer look at some of the ideas.

The idea of an **average** is a very common one in everyday life. Here are some examples:

Can you think of any more?

There are several different ways of finding an average: we are going to consider two of them, the **mean** and the **mode**.

Arithmetic Mean

Here is an example similar to the ones we met in **World of Mathematics**, Book 3.

Find the mean of 5, 8, 6, 7 and 4.

$$\text{Mean} = \frac{5 + 8 + 6 + 7 + 4}{5}$$

$$= \frac{30}{5}$$

$$= 6$$

The table below shows how we can work out the mean of a set of numbers when some of the numbers are repeated.

Mark out of 10	Number of pupils	
x	f	xf
3	4	12
4	3	12
5	5	25
7	5	35
8	3	24
totals	20	108

($xf \rightarrow x$ multiplied by f)

Mean mark: $\dfrac{108}{20} = \dfrac{54}{10} = 5.4$

A 1 Work out the mean (average) mark from the following results of a History test.

Mark out of 10	Number of pupils	
x	f	xf
2	1	
3	2	
4	4	
5	4	
6	5	
7	3	
8	1	
totals		

2 The following table shows the amounts spent by pupils at a school tuck shop at break on Monday morning.

Amount spent in pence	Number of pupils	
x	f	xf
2	7	
3	9	
4	8	
5	10	
6	5	
7	5	
8	3	
9	2	
10	1	
totals		

Find the average amount spent.

3 The table shows the amounts paid into the School Savings Bank by members of Forms 4 East and 4 West. Find which form had the higher average amount per pupil.

Form 4 East

Amount in pence	Number of pupils	
x	f	xf
5	2	
10	5	
15	5	
20	10	
25	4	
30	4	
totals		

Form 4 West

Amount in pence	Number of pupils	
x	f	xf
5	0	
10	3	
15	4	
20	5	
25	6	
30	2	
totals		

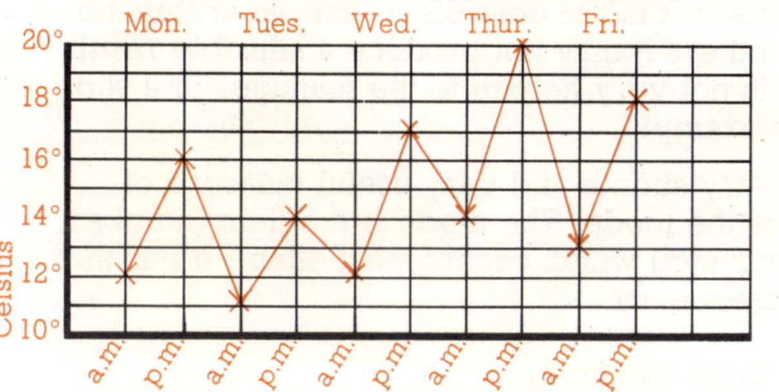

B

1. Peter and Ian each claimed he had the better batting record for the school cricket team. In five innings Peter had scored 10, 15, 2, 21 and 17. In six innings Ian had scored 12, 7, 14, 10, 18 and 5. Find which one was right by working out their mean scores.

2. On a seven-day touring holiday Mr. Watson and his family travelled the following distances: 100 km, 85 km, 10 km, 120 km, 79 km, 86 km and 80 km. What was the mean daily distance?

3. Six girls measured the length of a piece of material and obtained these results: 1.24 m, 1.26 m, 1.26 m, 1.25 m, 1.22 m and 1.27 cm. Find the mean length. (Can you see an easy method?)

4. In five training sessions David ran the 100 m in the following times: 11.7 s, 11.7 s, 11.8 s, 11.6 s and 11.4 s. What was the arithmetic mean for the five runs?

5. During one week the takings at a school tuck shop were: Monday £2.35, Tuesday £2.06, Wednesday £1.98, Thursday £1.62 and Friday £2.54. Find the average daily takings.

6. The temperatures taken during a school week are shown on the chart.

   a. Find the mean temperature for the mornings.
   b. Find the mean temperature for the afternoons.
   c. Make a copy of the above chart choosing your own scales.
   d. Draw lines on your chart to show the mean morning temperature and the mean afternoon temperature.

7. Divide yourselves into groups in one of the following ways:
   a. A class of boys or a class of girls: 4 groups.
   b. A mixed class: 2 groups of boys and 2 groups of girls.

   Find the following measurements for each member of the group and then calculate the mean for each measurement.

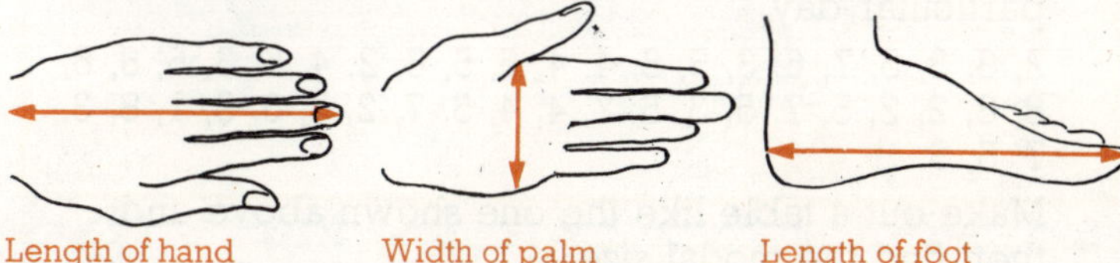

Mode or modal value

In some cases it is not possible to find an arithmetic mean, in others it may not produce a sensible result. Size 6.24 is not very helpful to the manager of a shoe shop, for example.

Another very simple and very useful measure of average is the **mode**. The mode is the item (number, size, colour etc.) which occurs most often — it has the greatest frequency.

Here is an example.

Men's shoes — Friday 12th January

Size	Record	Frequency
4					3
5	ⅢⅢ		6
6	ⅢⅢ |||	8
7	ⅢⅢ |||	8
8	ⅢⅢ ⅢⅢ ||	12
9	ⅢⅢ	5
10				2

In this case the mode is size 8.

C

1 Here is a list of women's shoe sizes sold on a particular day.

2, 3, 3, 5, 7, 6, 3, 5, 8, 4, 4, 5, 5, 8, 2, 4, 5, 3, 5, 6, 6, 8, 8, 2, 2, 5, 7, 5, 3, 5, 7, 4, 4, 5, 7, 2, 3, 6, 5, 1, 8, 3, 7, 5, 5.

Make out a table like the one shown above and then find the modal size.

2 Form 4 North obtained the following results from a survey of the colours of eyes among members of their form.

Colour	Record	Frequency
Brown	ⅢⅢ ⅢⅢ ||	
Blue	ⅢⅢ		
Green	ⅢⅢ ||	
Grey	||	
Hazel	ⅢⅢ	

a Find the frequency for each colour and the 'modal colour'.

b Carry out a similar survey among the members of your own class.

3 Scan the first three lines of any book (use as many different books as possible among the class) and make a tally of the number of times the vowels appear.

Find the 'modal vowel' and compare your results.

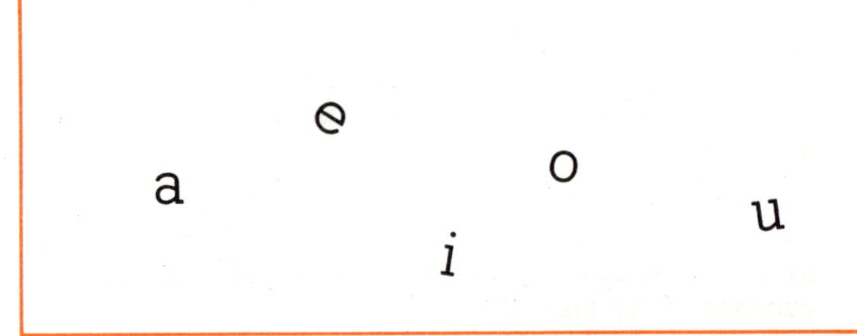

D Bar charts

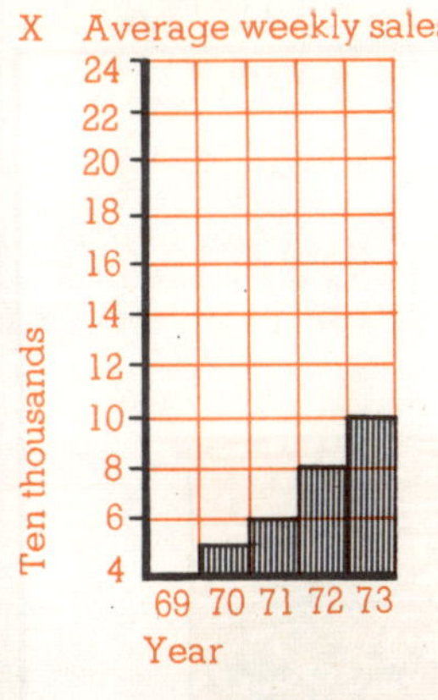

'Teenager' announces record growth in sales

X Average weekly sales

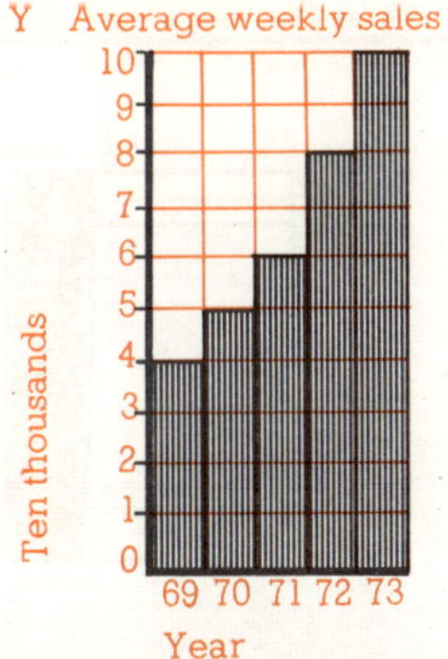

'Teenager' announces record growth in sales

Y Average weekly sales

1 Do the two bar charts give exactly the same information?

2 What were the average weekly sales in 1969?

3 What were the average weekly sales in 1973?

4 What was the increase in average weekly sales from 1969 to 1973?

5 Which chart, **X** or **Y**, seems to show the bigger sales?

6 Which chart, **X** or **Y**, seems to show the bigger increases in sales?

7 Which chart, **X** or **Y**, would the publishers choose to print in their magazine?

E The following table shows the number of books borrowed per year from a small public library.

Year	1968	1969	1970	1971	1972	1973
Thousands	7	8	10	14	17	20

1 On the same sheet of graph paper draw two bar charts to show the information in the table. Use the two different scales shown below.

Chart X
Horizontal axis: 1 cm to 1 year
Vertical axis: 2 mm to 1000 (starting at 5000)

Chart Y
Horizontal axis: 1 cm to 1 year
Vertical axis: 5 mm to 1000 (starting at 0)

2 The librarian is anxious to get money from the Council to buy more books. Which chart should he use to explain the increase in book borrowing? Explain why.

15

'Do it yourself'

Sooner or later, most of us have to tackle some job in the house or garden. Very often there is some measuring and working out of costs to be done.

In this chapter we are going to look at some examples of 'Do it yourself'.

A Tiling a bathroom wall

Before we start tiling we have to work out the number of tiles we need to buy. Here is an example.

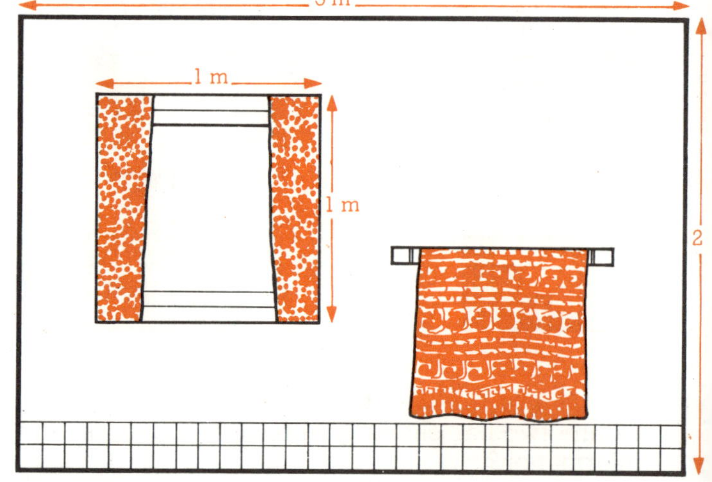

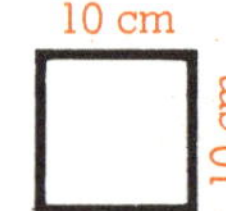

Area of the wall $= (3 \times 2)$ m²
$= 6$ m²

Area of the window $= (1 \times 1)$ m²
$= 1$ m²

Area to be tiled $= (6 - 1)$ m²
$= 5$ m²

Area of a tile $= (0.1 \times 0.1)$ m²
$= 0.01$ m²

No. of tiles needed $= (5 \div 0.01)$
$= (500 \div 1)$
$= 500$

Find the number of tiles needed to tile these two walls.

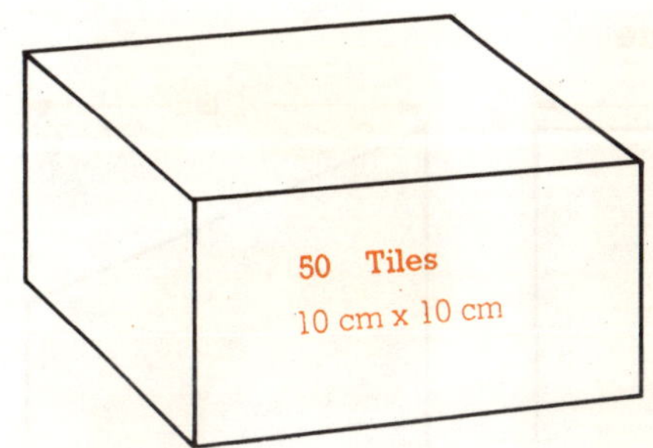

B   A box of tiles (each 10×10 cm) costs £1.20. Find the cost of the tiles needed for:

1   The wall shown for the example on p. 78.
2   The wall for question A 1
3   The wall for question A 2

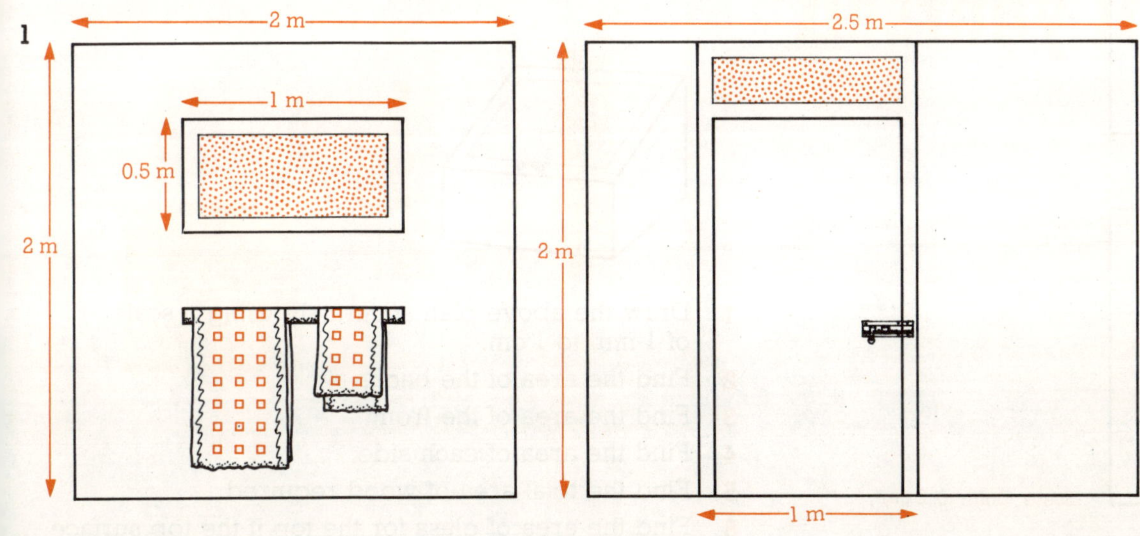

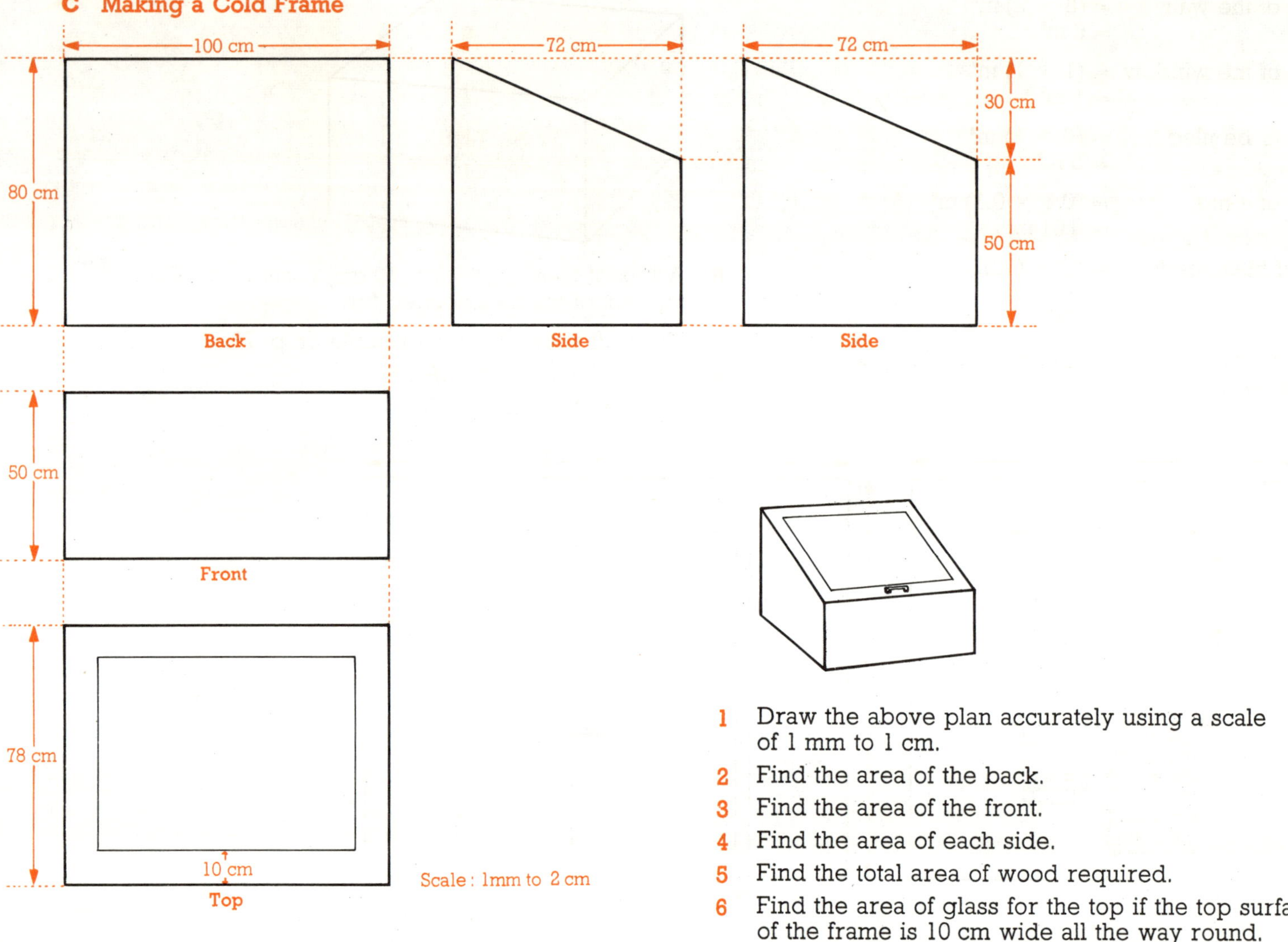

1 Draw the above plan accurately using a scale of 1 mm to 1 cm.

2 Find the area of the back.

3 Find the area of the front.

4 Find the area of each side.

5 Find the total area of wood required.

6 Find the area of glass for the top if the top surface of the frame is 10 cm wide all the way round.

1 Draw an accurate copy of the following plan.

Plan Scale : 1 cm to 1 m

2 Find the length of fencing needed to go right round the garden and the cost of the fencing.

3 Find the area of the path and the cost of laying the path at £1.40 a square metre.

4 Find the area of the lawn and amount of grass seed needed at the rate of 70 g a square metre.

5 How many kilogramme boxes of grass seed would it be necessary to buy and how much would they cost?

6 Find the total Hire-Purchase cost of the shed.

7 Find the total cost of the fencing, path, grass seed and garden shed.

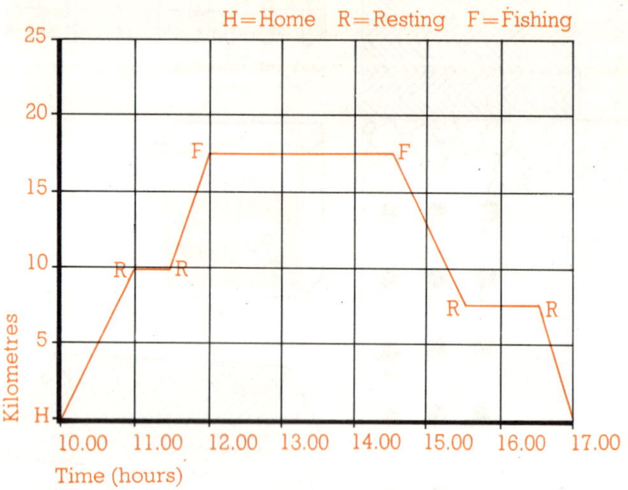

A The graph shows how Peter's time was spent on a day out cycling and fishing. Study the graph and then answer these questions.

1 How long did he cycle before his first rest?

2 What was his average speed between 10.00 and 11.00?

3 How long did he take for his first rest?

4 At what time did he start fishing?

5 How long did he stay fishing?

6 What was the time when he stopped for a rest on the way home?

7 How long did he take for his rest on the way home?

8 What was his average speed on the last part of his ride home?

9 How far did he cycle altogether?

10 How long was he away from home?

B Give the meter readings for January and April and then find how much gas had been used during that period.

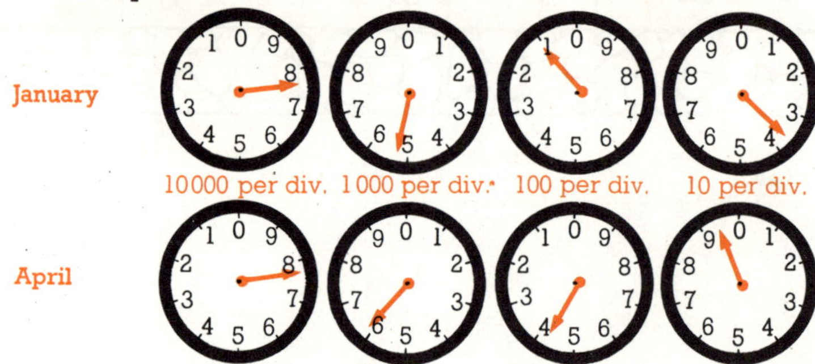

C Draw an accurate copy of this figure and then draw an 'envelope' design similar to those shown in Chapter 13; Assignment E.

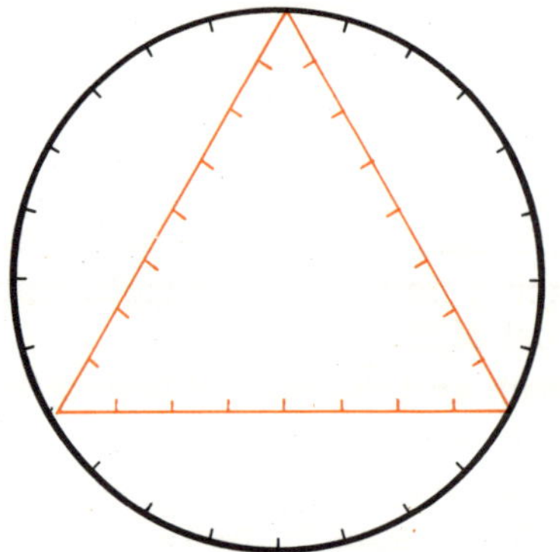

D When we are finding the mean value of a set of numbers it is sometimes possible to simplify the working. Here is an example.

2408 2406 2425 2433 2418

$$\text{Mean} = 2400 + \frac{8 + 6 + 25 + 33 + 18}{5}$$

$$= 2400 + \frac{90}{5}$$

$$= 2400 + 18$$

$$= 2418$$

Use the above method to find the mean of the following sets of numbers.

1 1502 1507 1503 1510 1517 1509
2 4937 4941 4988 4965 4974
3 10011 10007 10024 10003 10012 10004 10016
4 24120 24231 24085 24324

E 1 A survey on the popularity of football clubs produced the following results in one school. Write a list of the teams in order of popularity.

Newcastle	ЖЖ ЖЖ ЖЖ			
Liverpool	ЖЖ ЖЖ ЖЖ ЖЖ ЖЖ ЖЖ				
Leeds	ЖЖ ЖЖ ЖЖ ЖЖ		
Birmingham	ЖЖ			
Arsenal	ЖЖ ЖЖ ЖЖ		
Tottenham	ЖЖ ЖЖ				
Chelsea	ЖЖ ЖЖ ЖЖ ЖЖ			
Southampton	ЖЖ				

2 Here is a list of the amounts spent at a school tuck shop on one day. Find the frequency for each amount and the 'modal amount' spent.

5p, 3p, 3p, 7p, 3p, 8p, 2p, 3p, 5p, 7p, 4p, 8p, 8p, 6p, 9p, 10p, 5p, 6p, 6p, 3p, 4p, 8p, 7p, 7p, 3p, 4p, 4p, 5p, 8p, 6p, 3p, 4p, 4p, 5p, 8p, 8p, 9p, 2p, 2p, 5p, 3p, 3p, 5p, 5p, 7p.

F A gas fire burns 0.15 therms an hour.

Find

a the number of therms burned.

b the cost of the gas at 7p a therm if the gas fire is in use for:

1	3 hours	**3**	7 hours	**5**	15 hours
2	4 hours	**4**	10 hours	**6**	22 hours

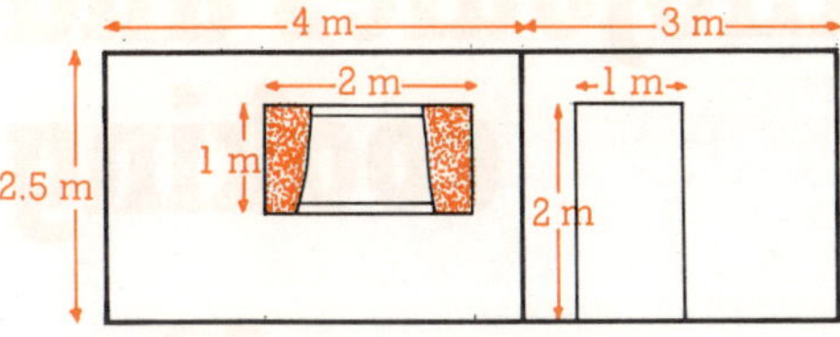

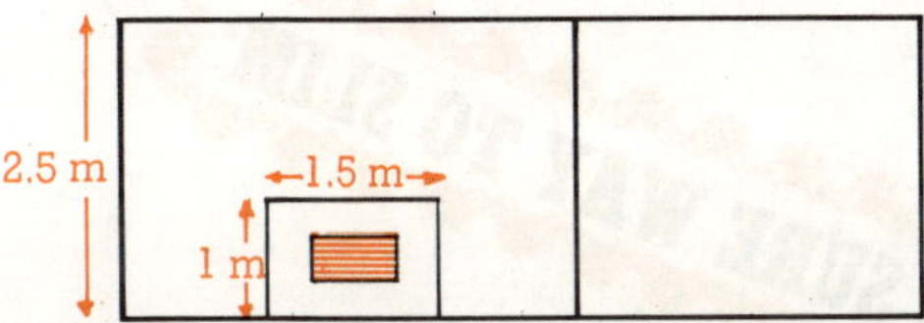

G On one brand of emulsion paint the instructions say that 1 litre will cover 10 m².

Work out how many litres will be needed to paint the walls of the above room taking away the areas of the window, door and fireplace.

16

Kilojoules and cooking

Almost everyday we see advertisements on television and in magazines telling us how to slim.

Very often these advertisements mention Calories or **kilojoules**. Since kilojoules are now important to many people here are some of the things we should know about them.

* A joule is a metric unit used to measure heat.

* A kilojoule is used as a measure of the energy value of food.

* Nearly all our food contains kilojoules but in varying amounts.

* If we take in more kilojoules a day than we need the extra kilojoules are turned into fat in our bodies.

Assignments

A Copy the following drawings and write the number of kilojoules per kilogramme under each one.

1 kg—39 000 kJ   1 kg—16 700 kJ   1 kg—13 600 kJ

1 kg—4000 kJ   1 kg—1900 kJ   1 kg—750 kJ

B Here is a list of foods giving the approximate number of kilojoules for each 100 grammes. Draw a table, as shown below, and then write the foods in order, according to the number of kilojoules they contain. Start with the largest number of kilojoules first.

Food	Kilojoules per 100 g
Beefburgers	5440

	Kilojoules		Kilojoules
Bacon	1670	Beefburgers	5440
Potatoes	420	Bananas	330
Scones	1550	Jam Roll	1670
Potato Crisps	2390	Custard	420
Apple Pie	840	Chicken	590
Swiss Roll	1760	Yogurt	210
Cheese	1800	Steamed Cod	330
Beef	1340	Sugar	1670
Milk	300	Plain biscuits	1880
Fresh Peas	250	Chocolate biscuits	2300
Cornflakes	1500	Carrots	85
Grapefruit	85	Rice Pudding	630
Butter	3770	Apples	210
Coconut Buns	2300	Fried Egg	1000
Bread	1050	Brussels Sprouts	130

C Use the list of foods in Assignment **B** to work out the number of kilojoules for each of the following snacks and meals.

1
Potatoe Crisps	100g
Apple Pie	50g
Glass of milk	150g

2
Steamed Cod	100g
Fresh Peas	50g
Potatoes	150g
Rice Pudding	70g

3
Cornflakes	60g
Milk	70g
Bread	80g
Butter	10g
Sugar	20g

4
Beef	100g
Carrots	100g
Potatoes	150g
Apple Pie	80g
Custard	50g

5
Bacon	80g
Fried Egg	50g
Bread	100g
Butter	20g
Chocolate Biscuit	30g
Milk	20g
Sugar	10g

6
Bread Roll	100g
Butter	10g
Cheese	50g
Milk	30g
Sugar	10g
Coconut Bun	40g

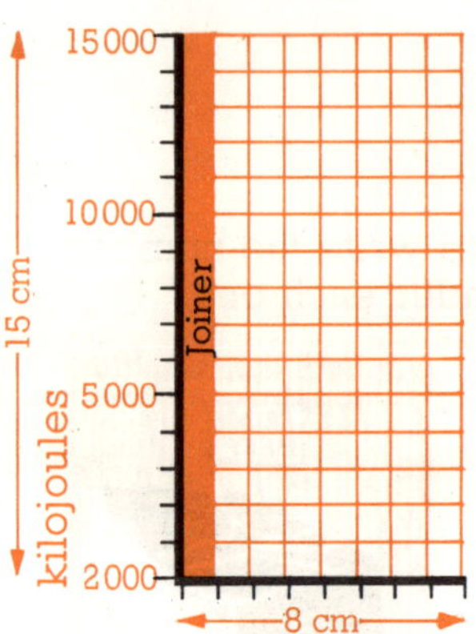

D

1   From the foods listed in Assignment **B**, choose the meal you would most enjoy, and then find how many kilojoules the meal would contain.

2   From the foods listed in Assignment **B**, make up a meal to be called 'The Slimmers Lunch', and then find out how many kilojoules it contains.

E   Recommended daily kilojoules

1   The following table shows the daily number of kilojoules recommended for various groups of people. Draw a bar chart to show this information. The next figure shows how to set out the chart and the measurements and scale to use.

Joiner	14 700
Male office worker	10 500
Female office worker	8 400
Housewife	10 500
Boy 16 to 20	14 200
Girl 16 to 20	10 500
Boy 13 to 15	13 400
Girl 13 to 15	11 700

Scale

1 cm to 1000 kilojoules
1 cm to each bar

Write a few sentences saying what you find interesting or unexpected about the information shown in your bar chart.

2   The following table shows the average weights for young men and women for various heights.

Height in cm	Women weight in kg	Men weight in kg
145	45	—
150	47	—
155	50	51
160	53	55
165	56	58
170	59	61
175	62	64
180	–	68
185	–	71

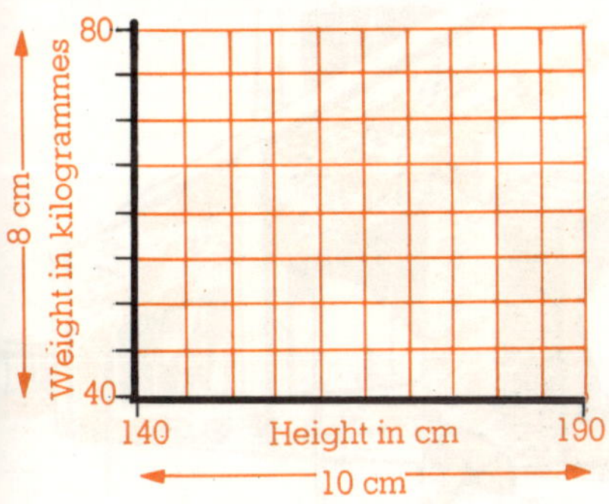

Draw a graph to show the information in the table.

The figure shows how to set out the graph.

You could use the following scales:
Horizontal axis (height): 1 cm to 5 cm.
Vertical axis (weight): 1 cm to 5 kg.

F For each of the following recipes:
1 Find the total weight in grammes.
2 Find the total number of kilojoules.
3 Find the time to start baking if the food is ready
   for a 12.30 lunch.

Fruit Pie

Cooking apples	500g
Lard	50g
Plain flour	100g (1465 kj)
Sugar	100g
$\frac{1}{2}$ teaspoon of salt	—
1 teaspoon ground cinnamon	—
Baking time: 40 minutes	

Custard Tart

Lard	50g
Plain flour	100g
2 eggs	60g (750 kj)
200 ml of milk	212g (600 kj)
1 tablespoon of sugar	20g
grated nutmeg	—
Baking time: 45 minutes	

Victoria Sandwich

Self raising flour	100g
Butter	100g
Caster sugar	100g
2 eggs	60g (750 kj)
3 tablespoons of jam	80g (920 kj)
Baking time: 20 minutes	
Cooling time: 30 minutes	

Bacon-and-Egg Pie

Plain flour	150g
Bacon	200g
Lard	100g
2 eggs	60g (750 kj)
2 tablespoons of milk	20g (60 kj)
$\frac{1}{2}$ teaspoon of salt	—
Baking time: 35 minutes	

Surveying

Whenever we travel about the country we are almost certain to see new roads, factories or housing estates being built.

Before buildings can start the land has to be surveyed, that is a plan or map of the area has to be drawn.

In this chapter we are going to look at some methods of surveying, and you can then carry out your own practical surveys.

Assignments

A Survey-lines

If we wish to draw a plan of the front of a building or the edge of a field we can use a survey-line. The figure shows a sketch of the front of a school building.

Notice that the lines on the sketch do not have to be drawn correctly to scale.

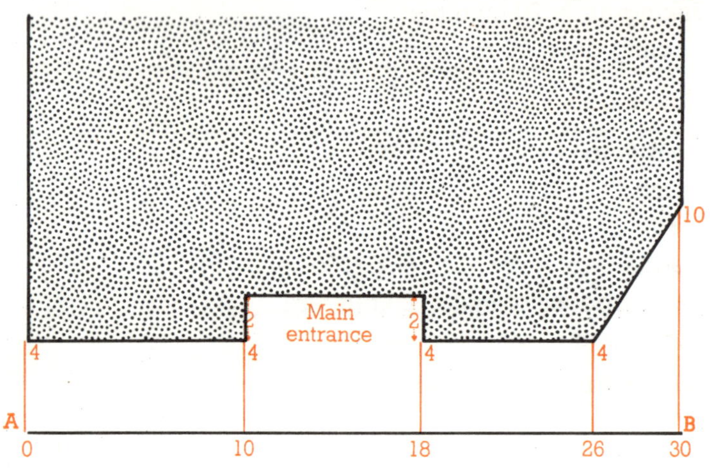

Here are the points to note:

* **AB** is the **survey-line** which is set out near the building.

* The coloured lines, known as **offsets**, are drawn at rightangles to the survey-line.

* The numbers along the survey-line show the distances in metres from station **A**.

* The numbers at the end of each offset show the distance of each corner of the building from the survey-line.

* The measurements shown by arrowed lines give the lengths of parts of the wall.

1 Draw an accurate plan of the front of the school building. Start by drawing the survey-line and use a setsquare for the offsets. Use a scale of 1 cm to 2 m.

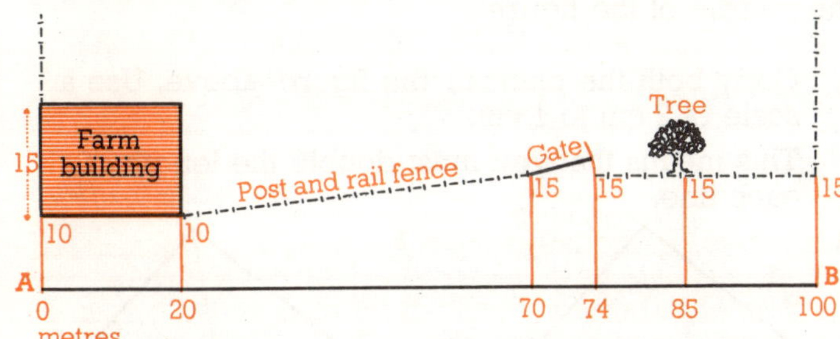

2 The figure above shows a survey of one side of a field. Draw an accurate plan of the side of the field using a scale of 1 cm to 5 m.

The figure to the right shows the best way to set out a **field-book** when a survey-line is used to draw a plan.

First a rough sketch of the area is drawn and then a column is ruled out to represent the survey-line.

Notice that the station from which the survey starts is put at the bottom of the page and that it is not necessary to draw in the offsets.

3 Use the field-book sketch shown in the figure below to draw an accurate plan. Work to a scale of 1 cm to 5 m.

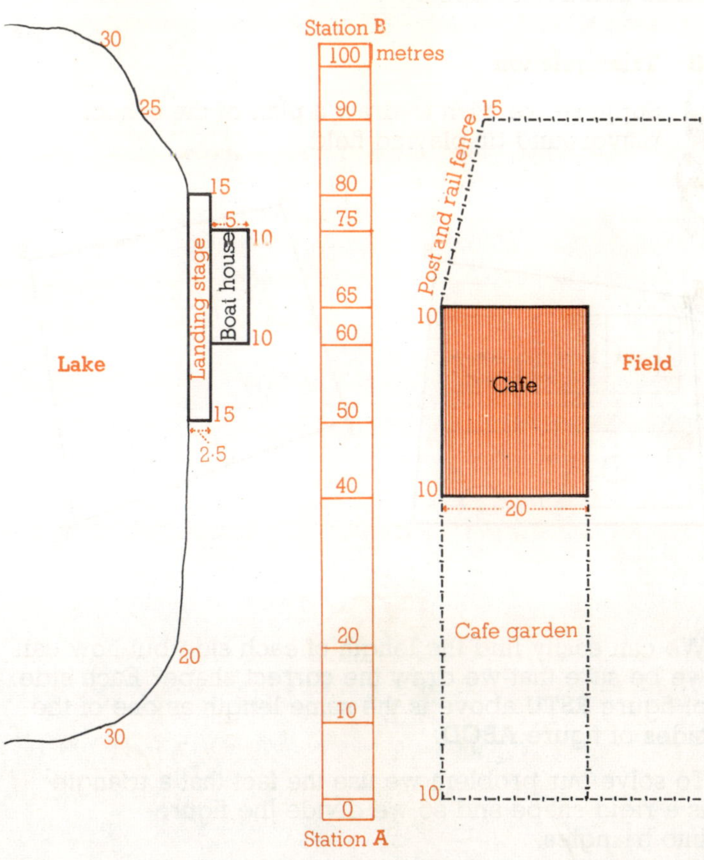

Use the survey-line method to carry out a survey in the vicinity of your school or playing field.

You will need at least four poles, known as ranging rods, two tapes marked in metres and centimetres, a number of arrows (a kind of spike), a piece of chalk and a field-book.

B Triangulation

Suppose we wish to draw a plan of the school playground or playing field.

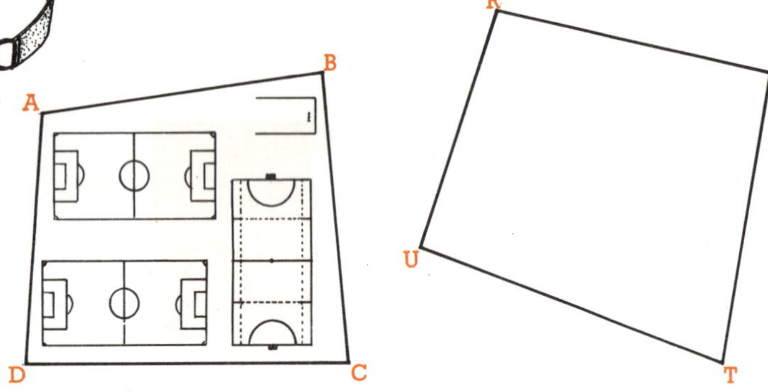

We can easily find the length of each side but how can we be sure that we draw the correct shape? Each side of figure **RSTU** above is the same length as one of the sides of figure **ABCD**.

To solve our problem we use the fact that a triangle is a rigid shape and so we divide the figure into triangles.

Here are two ways in which this can be done.

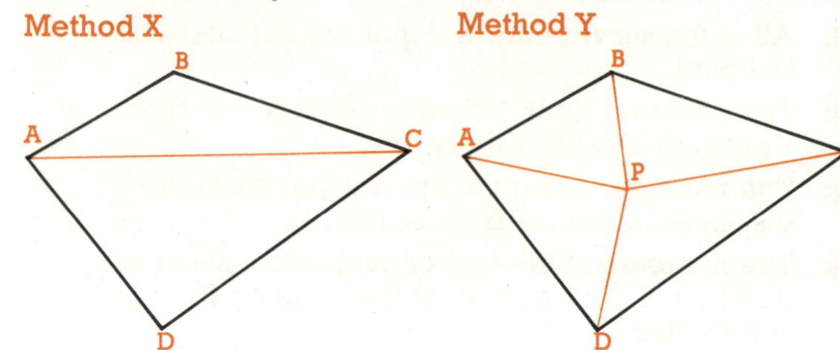

For Method X, which is suitable for small areas, we need to find the lengths of **AB**, **BC**, **CD**, **DA** and **AC**.

For Method Y, which is suitable for larger areas, we need to find the lengths of **AB**, **BC**, **CD**, **DA**, **PA**, **PB**, **PC** and **PD**. The point **P** should be somewhere near the middle of the figure.

1   Copy both the plans in the figure above. Use a scale of 2 cm to 1 cm.

   This means that you must double the length of each line.

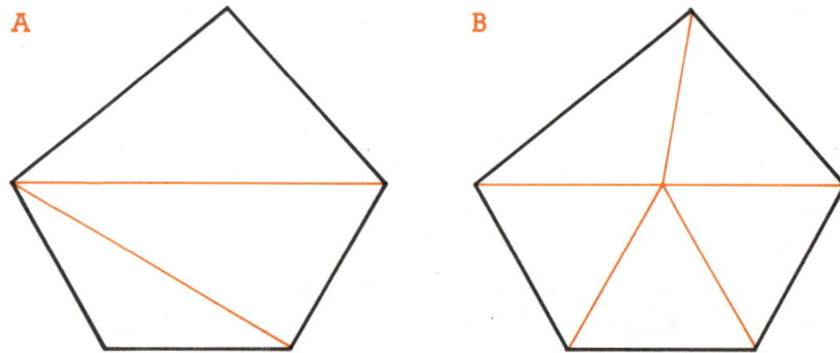

2   Copy both the plans above. Use a scale of 2 cm to 1 cm.

Use the triangulation method to survey your own playground, playing field or some similar area.

You will need a number of ranging rods, a tape, arrows or chalk, and a field-book.

Use a ranging rod to mark the point **P** somewhere in the middle of the area that you are surveying.

C Triangulation using angles

The figure shows how to survey an area by dividing it into triangles and measuring the angle of each corner, using 360-degree compass bearings.

* A point **P** is selected in the middle of the area.
* The distance from **P** to each corner of the field is measured.
* A compass bearing is taken from **P** to each corner of the area.

Scale: 1 cm to 10 m

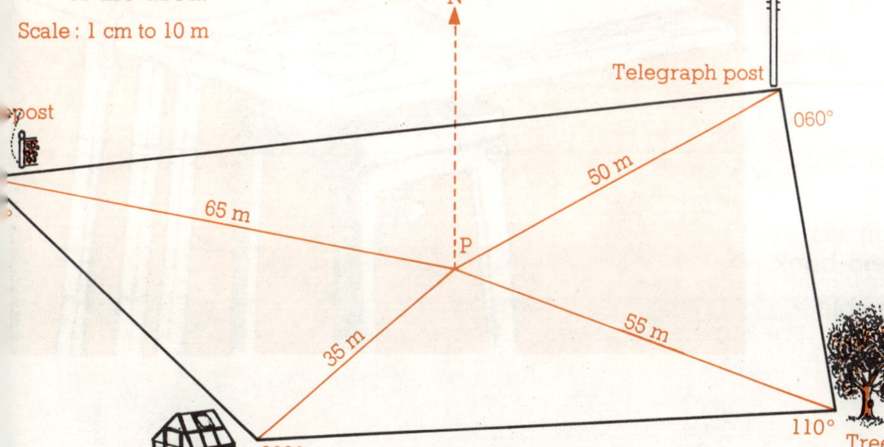

1 Draw accurate copies of these two figures.

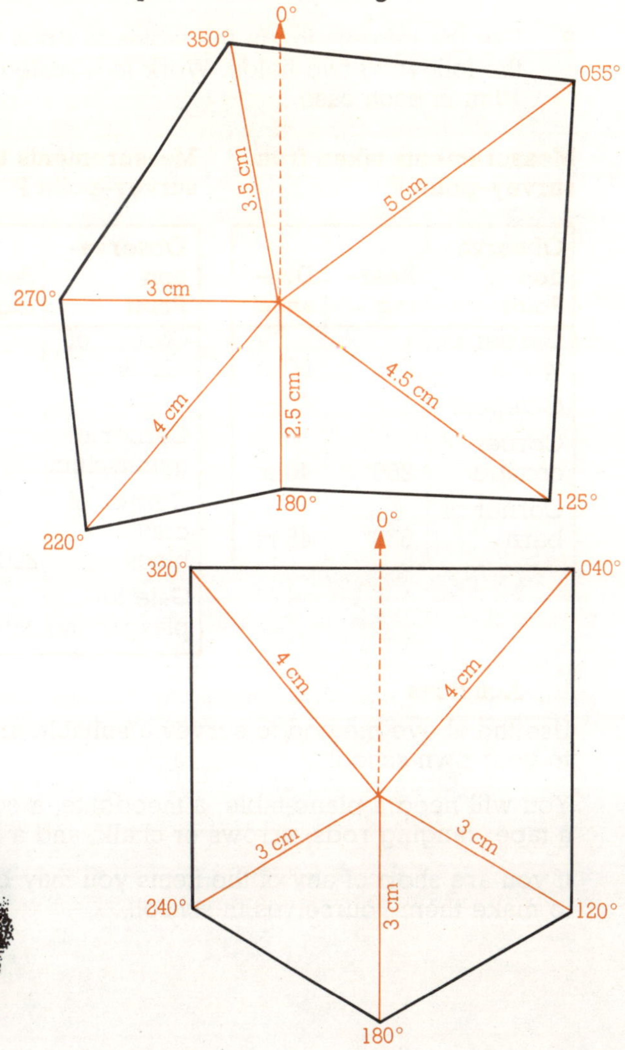

2 Copy the plan of the field on p. 91.

3 Use the information in the tables to draw plans of the following two fields. Work to a scale of 1 cm to 10 m in each case.

Measurements taken from survey–point P

Observation Point	Bearing	Distance
Corner of wood	045°	60 m
Gatepost	170°	50 m
Corner of cottage	260°	40 m
Corner of barn	330°	45 m

Measurements taken from survey–point P

Observation Point	Bearing	Distance
Corner of tennis court	045°	50 m
Corner of gymnasium	135°	50 m
Corner of craft block	220°	55 m
Gate to playground	285°	60 m

Activities

Use the above method to survey a suitable area near to your own school.

You will need a plane-table, a theodolite, a compass, a tape, ranging rods, arrows or chalk, and a field-book.

If you are short of any of the items you may be able to make them yourselves in school.

18 Rates

Some people live in towns others live in the country but all of us need such things as water, roads and schools. These and many other services are essential if we are to live in comfortable, healthy and pleasant surroundings.

The services are provided by local councils and a great part of the cost is met from the rates.

Today local government plays a very important part in the lives of almost every one of us.

Assignments

A Copy and complete the following sentences.

1 What are rates for?

Rates are a means of raising money to provide for local services. These services include such things as education, housing, libraries ▰▰▰▰▰▰▰

R.V. £120

R.V. £350

R.V. £500

R.V. £2000   =

2   Who has to pay rates?

Rates are paid by the occupiers of properties such as houses, shops, ▮▮▮▮▮▮▮▮

3   How are the rates worked out?

a   Each property is given a rateable value (R.V.).

b   A rate in the £1 is fixed for each year.

c   The amount of rates to be paid is the 'rate in the £1' multiplied by the R.V.

B   The rateable value of each property is decided by the local Valuation Officer.

In the case of a house the R.V. depends on the amount of land, the number and size of the rooms, and the amenities in the house such as bathrooms and central heating.

Some examples of rateable values are given in the pictures above.

1   Draw the four different properties shown above and write the R.V. of each one underneath.

2   Write out a list of buildings in your area and make up a R.V. for each one. You may be able to find out the correct R.V.s from a list in your council offices.

C

1   Work out the rates to be paid for each of the properties shown in Assignment B if the rate in the £1 is 50p. This means that for each £1 of rateable value you have to pay 50p.

Copy the first example which has been done for you.

Small house R.V. £120

Rate in the £ : 50p   rates to be paid : 120 x 50p = 6000p
= £60.00

2 Write down the following examples and work out the rates to be paid in each case.

	R.V.	Rate in the £1	Rates to be paid
a	£200	60p	200 × 60p = £120.00
b	£200	70p	200 × 70p = £140.00
c	£100	50p	100 × 50p =
d	£100	60p	100 × 60p =
e	£300	40p	
f	£300	41p	
g	£250	30p	
h	£400	54p	
i	£420	70p	
j	£535	40p	
k	£560	60p	
l	£1 000	68p	
m	£1 000	72p	
n	£2 000	72p	
o	£2 340	44p	

D Each council has a Treasurer who is responsible for the rates. He knows the total R.V. for his area and so he can work out the amount he will collect for each 1p in the £1 on the rates.

Here are some examples :

Total R.V.	Rate in the £1	Amount collected
£1 000 000	1p	1 000 000 × 1p = 1 000 000p = £10 000
£1 000 000	2p	1 000 000 × 2p = 2 000 000p = £20 000
£500 000	1p	500 000 × 1p = 500 000p = £5 000
£500 000	3p	500 000 × 3p = £15 000

Now copy and complete the following table.

	Total R.V.	Rate in the £1	Amount collected
1	£200 000	1p	
2	£200 000	2p	
3	£200 000	3p	
4	£500 000	2p	
5	£500 000	4p	
6	£800 000	1p	
7	£800 000	2p	
8	£1 000 000	3p	
9	£1 000 000	4p	
10	£1 000 000	10p	
11	£2 000 000	1p	
12	£2 000 000	5p	

E Here is a simplified example of part of a statement which appears on the back of a rate demand note.

Statement of expenses

The estimated product of a penny rate is £40 000.

	p
Education	50
Public Health	6
Social Services	7
Housing	9
Fire Services	2
Highways and Bridges	8
Parks and Gardens	3
Libraries and Museums	4

1 Find the amount spent on each service shown on the statement.
Copy the first example which has been done for you.

Education

The product of 1p in the £1 is £40 000
The product of 50p in the £1 is £40 000 × 50
i.e. £2 000 000

2 In the above example the product of a penny rate is £40 000. What is the total rateable value of all the property in the area?

19 Computers

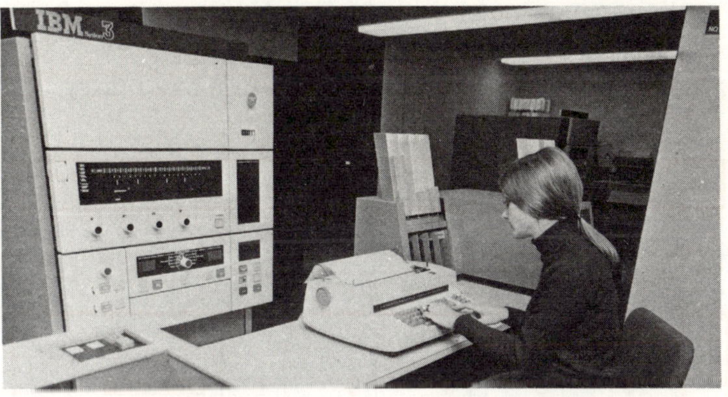

Computers are becoming more and more important in everyday life. You have probably seen a number of television programmes which have shown some of the ways in which they are being used.

Although few of us will ever help to build or operate a computer, many people are finding that computers are helping in the work they have to do.

Assignments

A Give as many examples as possible of the ways in which computers are being used. Here are five applications to start your list.

- aircraft navigation
- working out wages
- keeping medical records
- teaching
- ordering goods for supermarkets

B Although we have seen that computers have a great variety of uses, most of them are 'digital computers' which means that they use numbers.

You may remember that computers use a base-2 system of numbers, called the binary system.

Here are the place values of the first five binary digits.

sixteen	eight	four	two	one

Copy the following table and give the decimal equivalents of the binary numbers

| | Binary | | | | | Decimal |
	16	8	4	2	1	
1				1	1	3
2			1	0	0	
3			1	0	1	
4			1	1	1	
5		1	0	0	1	
6		1	0	1	1	
7		1	1	0	1	
8	1	0	0	0	1	
9		1	1	1	1	
10	1	0	1	0	1	
11	1	1	0	1	0	
12	1	1	0	1	1	
13	1	0	1	1	1	
14	1	1	1	0	1	
15	1	1	1	1	1	

C Copy the following table and then find the binary numbers equivalent to the decimal numbers. Squared paper would be very useful for this assignment.

| | Decimal | Binary | | | | |
		16	8	4	2	1
1	27	1	1	0	1	1
2	6			1	1	0
3	9					
4	12					
5	17					
6	23					
7	11					
8	25					
9	30					
10	24					
11	18					
12	13					
13	26					
14	31					

D Binary numbers can be represented in many different ways. Here is one device which consists of a row of bulbs. A bulb which is 'off' represents a 0, a bulb which is 'on' represents a 1. Here is an example.

The number represented by these bulbs is:

$$1\ 0\ 1\ 0\ 1\ 1 \longleftrightarrow 4\ 3$$

Base two   Base ten

Find (a) the binary number and (b) the decimal number represented by each of these sets of bulbs.

1

2

3

4

5

6

7

8

E One method of feeding information into a computer is by using punched tape. The information is punched into the tape using binary numbers.

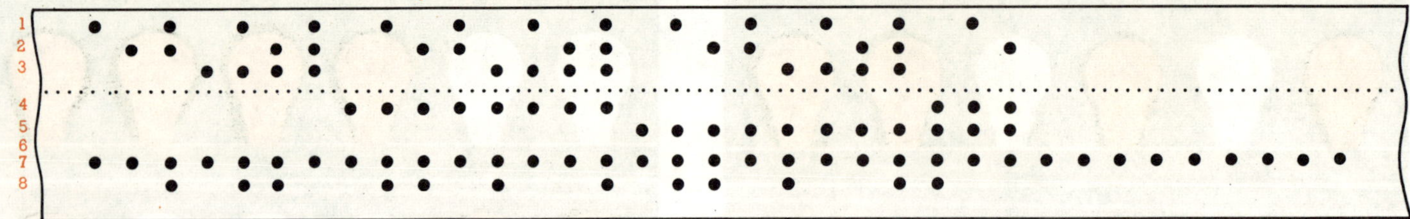

The most common form of tape, shown above, uses eight tracks of coding holes. The track of smaller holes is to feed the tape through the computer reader.

Look at the first five tracks of holes. If we take a hole to represent 1 and a blank to represent 0 we see that the letters of the alphabet are represented by the binary numbers one to twenty-six.

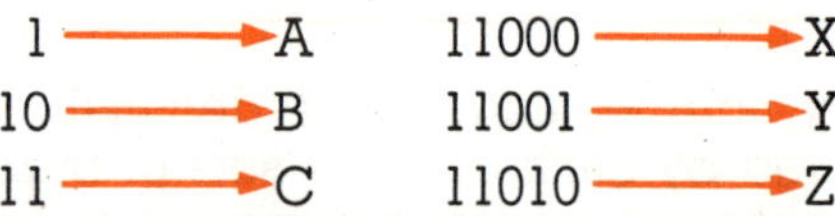

Use the binary code to decipher the messages on the following tapes.

1 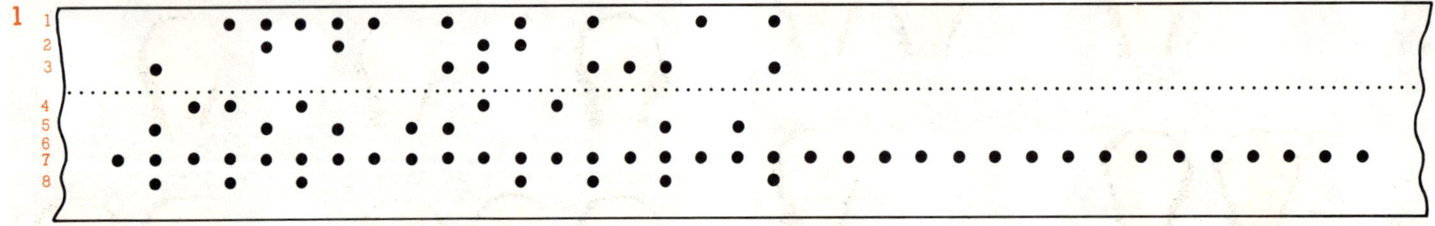

2 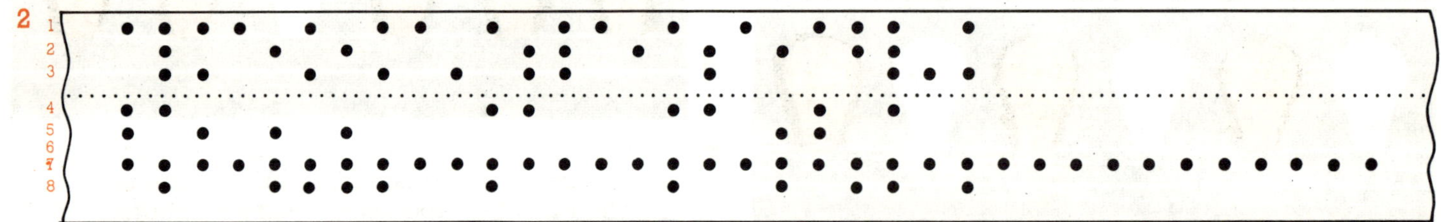

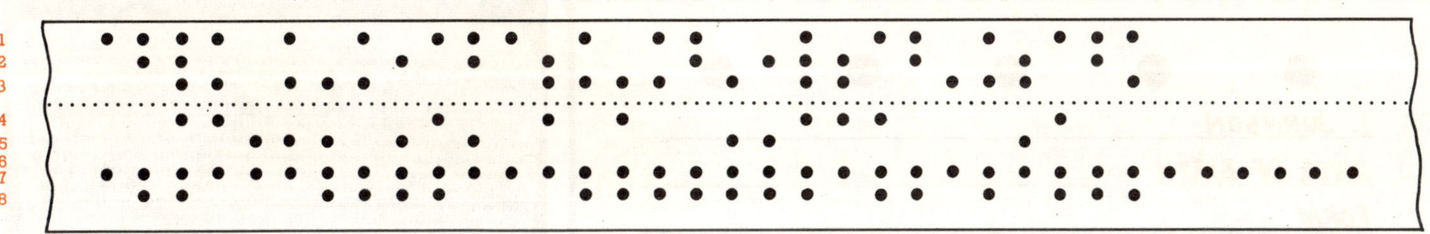

Punched cards

A binary code can be used to arrange cards in order.

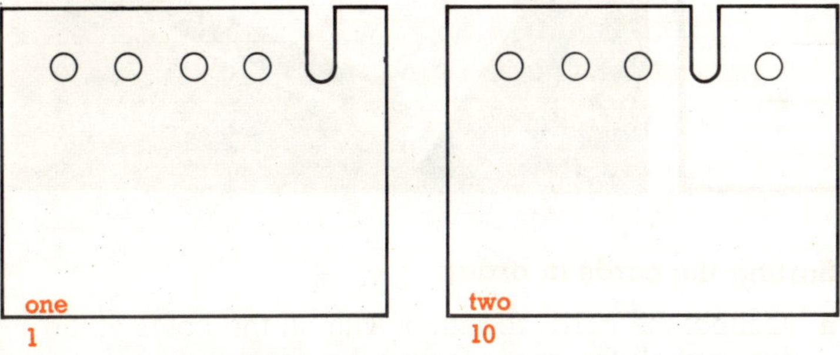

The binary numbers are represented by slots and holes as shown in the cards.

A hole represents a 0, a slot represents a 1.
The binary numbers represented by the slots and holes are shown underneath the card in each case.

Carry out the following instructions to make a card index system for your own class or group.

1 On the blackboard write a list of the names of your group in alphabetical order.

2 Number the list of names, first in decimal and then in binary.

Alan, J.	1	1
Beckett, W.	2	10
Bridgewood, E.	3	11
Chapman, J.	4	100

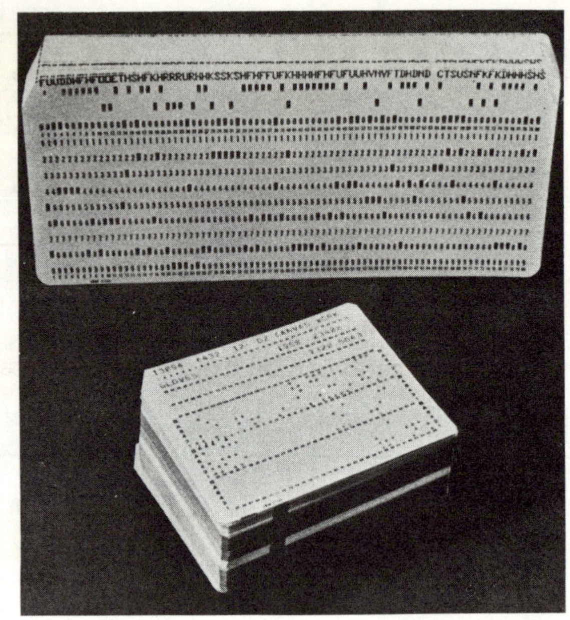

3 Cut out your own card, measuring 12 cm by 8 cm.

4 Punch 5 holes in the top of your card in the positions shown. The measuring must be exact.

5 Write your name on your card and then cut slots to represent the binary number that you have been given.

6 Decide what information to write about yourselves on your cards.

Sorting the cards in order

a Collect the cards together with all the holes at the top and all the cards facing the correct way.

b Place a compass point through the right-hand holes and lift the cards on the compass clear of the rest.

c Put the cards you have taken out at the back of the pack.

d Repeat operations **b** and **c** using holes 2, 3, 4 and 5.

If you have carried out the instructions correctly your cards should now be in the correct order.

20 Constructions

Geometrical constructions are frequently used by draughtsmen for drawing plans, and they play an important part in design work generally.

Remember that neatness and accuracy are essential when you practise these constructions.

A   A triangle inside a circle

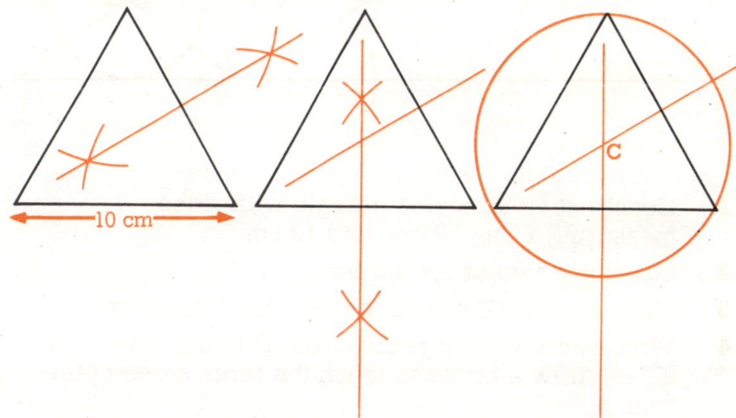

1   Construct an equilateral triangle with sides measuring 10 cm.

2   Bisect any two of the sides of the triangle.

3   Mark a point **C** where the two bisectors cross.

4   With centre **C**, and radius from **C** to any corner of the triangle, draw a circle.

Use your construction as the basis of a design.

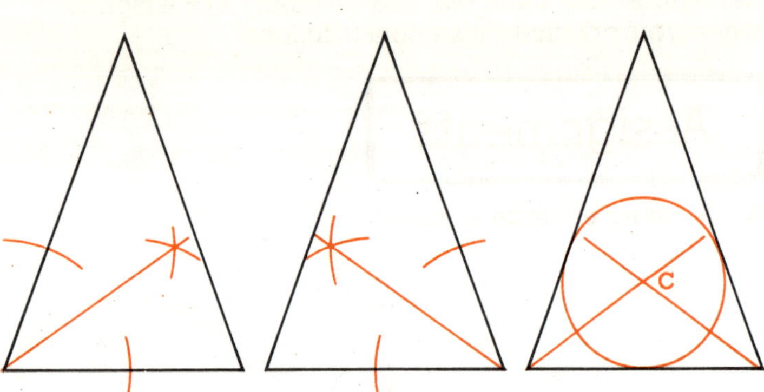

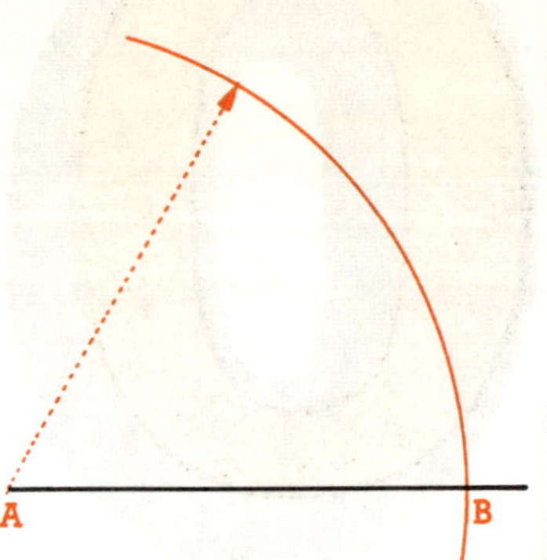

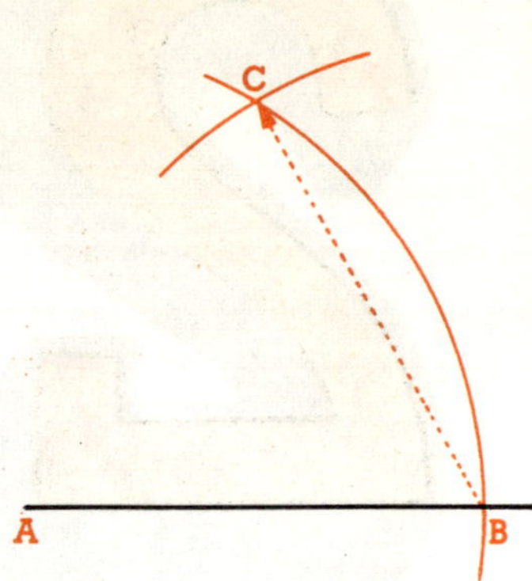

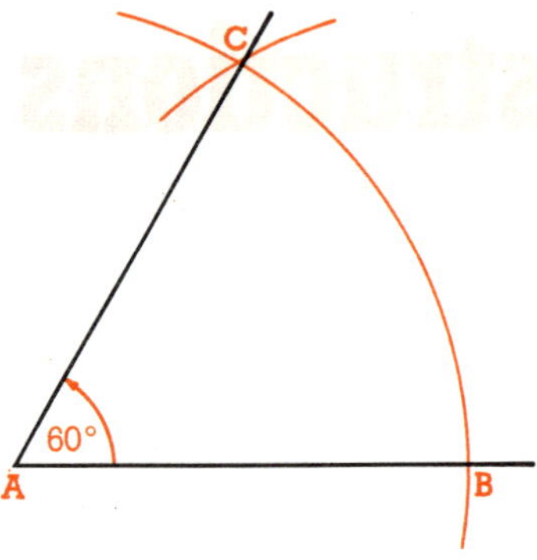

1 Construct an isosceles triangle with sides measuring 8 cm, 12 cm, and 12 cm.

2 Bisect any two of the angles.

3 Mark a point **C**, where the two bisectors cross.

4 With centre **C**, and radius from **C** to any one of the sides, draw a circle to touch the three sides of the triangle.

Do constructions **A** and **B** again. This time use a triangle with sides measuring 9 cm, 8 cm and 7 cm.

1 Draw a line any convenient length.

2 With compasses set at a convenient radius (say 4 cm), and centre at **A**, draw an arc as shown above.

3 With centre **B**, and the same radius, draw an arc to cut the first arc at **C**.

4 Draw a line from **A** through **C**.
Measure angle **CAB** and check that it is 60°.

D 1 Square Pyramid

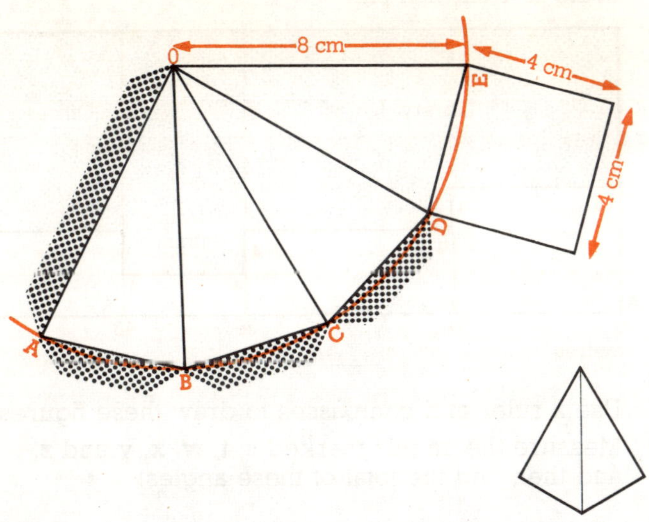

Look at the figure above and then draw the net, or plan, of a square pyramid on thin card as follows.

1 With centre 0 draw a semicircle radius 8 cm.
2 Mark a point **A** on the semicircle.
3 With radius 4 cm, and centre **A**, step off four points, **B**, **C**, **D** and **E**, round the semicircle.
4 Draw the lines: **AO**, **BO**, **CO**, **DO**, **EO**, **AB**, **BC**, **CD** and **DE**.
5 Draw a square with **DE** as one side.
6 Draw the flaps as shown.

Cut out the plan and assemble the pyramid.

2 Regular Octahedron

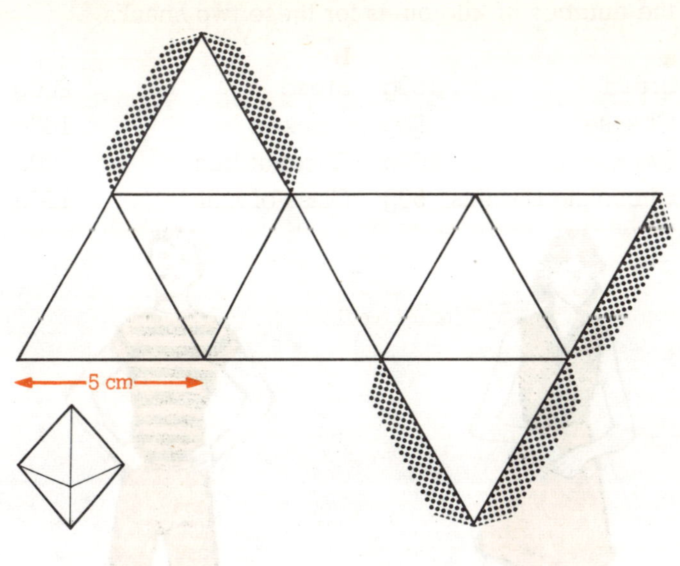

1 Practise drawing the net of a regular octahedron shown in the figure above. The net is made up of eight equilateral triangles.
2 Draw the net on thin card and then make a model of the octahedron.

4th Review

A Refer to the list of foods on page 85 and work out the number of kilojoules for these two snacks.

	a		**b**	
1	Bread	150g	Bread	200g
	Cheese	50g	Chicken	100g
	Swiss Roll	100g	Coconut bun	100g
	Chocolate biscuits	50g	Glass of milk	150g

2 Make up meals for a day for a girl so that she has about the right number of kilojoules. You will need to refer to the list of foods on page 85.

3 Repeat question **2** for a boy.

B

1 Use this sketch and measurements to draw an accurate plan of the front of the building. Use a scale of 1 cm to 4 cm.

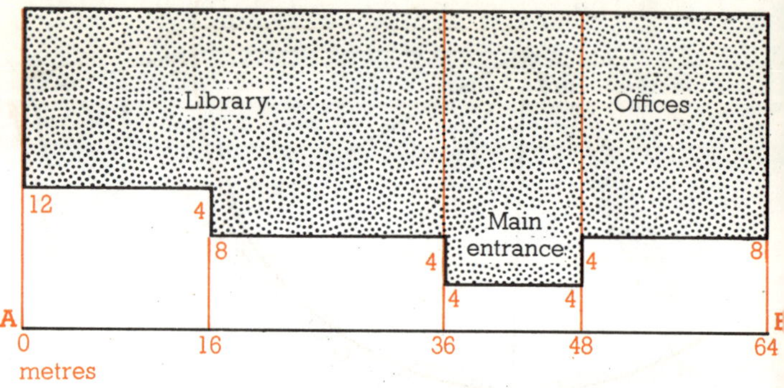

2
a Use a ruler and compasses to draw these figures.

b Measure the angles marked **s**, **t**, **w**, **x**, **y** and **z**, and then find the total of these angles.

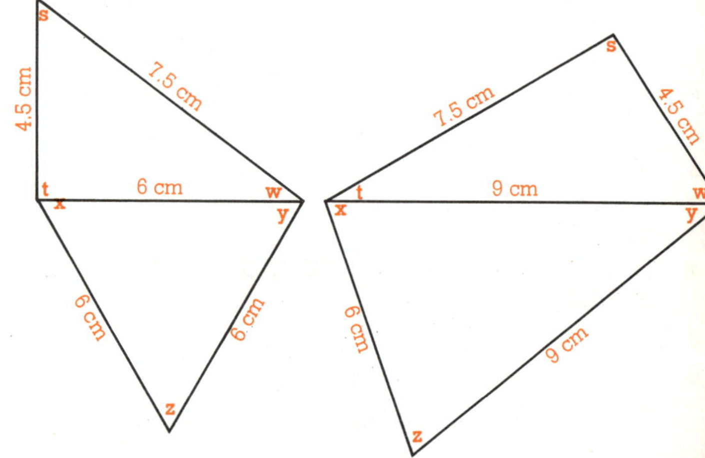

R.V. £20

R.V. £230

R.V. £645

R.V. £750

C

1 Work out the rates to be paid on these buildings if the rate in the pound is 75p.

2 The total Rateable Value of a town is £2 000 000. How much extra will the Borough Treasurer collect if he increases the rate in the £1 by a penny?

3 The total Rateable Value of a town is £3 500 000. How much extra will the Borough Treasurer collect if he increases the rate in the £1 by twopence?

4 If an extra penny in the £1 on the rates brings in £100 000 what is the total Rateable Value of the town?

D Give the binary numbers equivalent to these decimal numbers.

1	2	**4**	9	**7**	16	**10**	23
2	3	**5**	10	**8**	17	**11**	31
3	7	**6**	12	**9**	20	**12**	34

E Give the decimal numbers equivalent to these binary numbers.

1	10	**4**	111	**7**	1101	**10**	10010
2	101	**5**	1000	**8**	1111	**11**	11000
3	110	**6**	1010	**9**	10000	**12**	10101

F

1 Draw a copy of this tape on a piece of plain paper.

2 Refer to the binary code for the alphabet on page 100 and then draw and punch holes to represent **your own name**.

 Use a ruler and compasses to draw this figure
and then make it the basis of a pattern.

Draw the triangle first and choose your own
measurements.

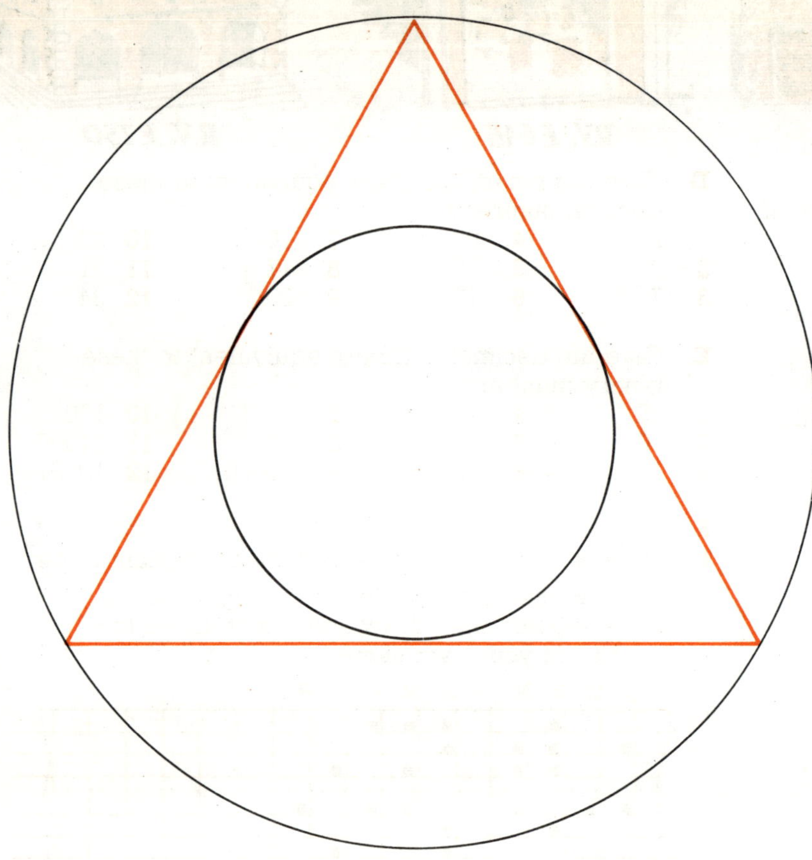